Springer Theses

Recognizing Outstanding Ph.D. Research

Aims and Scope

The series "Springer Theses" brings together a selection of the very best Ph.D. theses from around the world and across the physical sciences. Nominated and endorsed by two recognized specialists, each published volume has been selected for its scientific excellence and the high impact of its contents for the pertinent field of research. For greater accessibility to non-specialists, the published versions include an extended introduction, as well as a foreword by the student's supervisor explaining the special relevance of the work for the field. As a whole, the series will provide a valuable resource both for newcomers to the research fields described, and for other scientists seeking detailed background information on special questions. Finally, it provides an accredited documentation of the valuable contributions made by today's younger generation of scientists.

Theses are accepted into the series by invited nomination only and must fulfill all of the following criteria

- They must be written in good English.
- The topic should fall within the confines of Chemistry, Physics, Earth Sciences, Engineering and related interdisciplinary fields such as Materials, Nanoscience, Chemical Engineering, Complex Systems and Biophysics.
- The work reported in the thesis must represent a significant scientific advance.
- If the thesis includes previously published material, permission to reproduce this must be gained from the respective copyright holder.
- They must have been examined and passed during the 12 months prior to nomination.
- Each thesis should include a foreword by the supervisor outlining the significance of its content.
- The theses should have a clearly defined structure including an introduction accessible to scientists not expert in that particular field.

More information about this series at http://www.springer.com/series/8790

Shilei Zhang

Chiral and Topological Nature of Magnetic Skyrmions

Doctoral Thesis accepted by
the University of Oxford, Oxford, UK

 Springer

Author
Dr. Shilei Zhang
Clarendon Laboratory,
  Department of Physics
University of Oxford
Oxford, UK

Supervisor
Prof. Thorsten Hesjedal
University of Oxford
Oxford, UK

ISSN 2190-5053   ISSN 2190-5061 (electronic)
Springer Theses
ISBN 978-3-030-07472-2   ISBN 978-3-319-98252-6 (eBook)
https://doi.org/10.1007/978-3-319-98252-6

This Springer imprint is published by the registered company Springer Nature Switzerland AG
The registered company address is: Gewerbestrasse 11, 6330 Cham, Switzerland

To Valen:

The resonance with you scatters through every corner of my phase space.

Supervisor's Foreword

Magnetic skyrmions are swirls in a magnetic spin system, analogous to the skyrmion particle originally described in the context of pion fields. In their internal structure, constituent spins point in all the directions wrapping a sphere, which is mathematically described by a distinct topological index and can be a source of emergent phenomena such as the topological Hall effect. Skyrmions are observed in non-centrosymmetric B20 compounds in which Dzyaloshinskii–Moriya (DM) interaction plays a role. The competition between symmetric and antisymmetric exchange interactions gives rise to long-range ordered modulations, which are manifested by the hexagonal skyrmion lattice at specific temperatures and fields. Magnetic skyrmion materials hold the promise of rich novel physics and have the great advantage of a robust topological magnetic structure, which makes them stable against the superparamagnetic effect and are therefore a candidate for the next generation of spintronic memory devices.

Dr. Shilei Zhang's thesis aims for a comprehensive introduction to this fascinating field of skyrmions, building up the theory of magnetic skyrmions step by step from the atomic level (Chap. 1). An important focus of this book is on the experimental x-ray scattering techniques, in particular resonant elastic x-ray scattering (REXS), which have been used to study skyrmions across all length scales. Chapter 2 introduces the scattering theory, starting from single electrons and extending to all relevant ordered phases, as well as the main scattering techniques, i.e. the reflection and the grazing incidence geometry. Using the unique capabilities of REXS, such as element selectivity and sensitivity to both charge and spin ordering, it is possible to unambiguously resolve all magnetic phases. This includes the clear discrimination of the conical, ferrimagnetic and paramagnetic phases, which are indistinguishable in both small angle neutron scattering and Lorentz transmission electron microscopy.

Cu_2OSeO_3 is a highly unique magnetoelectric material displaying a diverse range of interesting magnetic textures and properties. Until Shilei's groundbreaking work, only single-domain, long-range ordered skyrmion lattices had been known to exist in a small pocket in the temperature-magnetic field phase diagram, and their occurrence is believed to be a universal feature of all B20-type skyrmion-carrying materials. In Chap. 3, the discovery of a new skyrmion state that consists of multiple, in-plane rotated hexagonal skyrmion lattices is presented. This multidomain skyrmion state is obtained for magnetic fields tilted away from the crystalline [001] axis, or in general, for fields not aligned along major crystallographic axes. The origin of the multidomain state can be understood in the framework of competing anisotropies and magnetoelectric effects.

The key quantity that defines the topological properties of a magnetic skyrmion is the winding number in real-space. The common, indirect approach to determine the winding number is to experimentally obtain an image of the magnetisation distribution, and compare it with simulations of the contrast of various, possible magnetisation patterns. However, it is of crucial importance that the topological quantity can be unambiguously determined experimentally, as this enables the profound understanding of the physical systems. In Chap. 4, Shilei presents a new general physical principle that allows direct access to the topological property of materials through light-matter interaction. It is naturally encoded in the polarisation dependence of the resonant x-ray scattering process, which is a universal process applicable to many different materials and systems. In particular, the circular dichroism and the linear polarisation dependence, the so-called polarisation-azimuthal maps, turn REXS into a powerful probe of topology. He introduces the topology determination principle and presents analytical solutions, numerical calculations, as well as experimental data, which show that this novel technique is exclusively sensitive to the winding number.

One of the major challenges in skyrmion science is the determination of the three-dimensional magnetisation distribution. It has been predicted already in 2013 that a twisted skyrmion surface state should exist in thin-film helimagnets, however, such a surface state has never been directly observed. The key parameter that identifies such a twisted state is the helicity angle of the surface skyrmions, which remained elusive for all magnetic characterisation techniques so far. Only the two extreme types of so-called Néel- and Bloch-type skyrmions have been recognised and experimentally confirmed to exist. Their helicity angle χ is $0°$ and $90°$, respectively. In Chap. 5, direct experimental evidence and a systematic study of the twisted skyrmion surface state with a non-trivial helicity angle is presented. Using circular dichroism in resonant elastic x-ray scattering, the helicity angle of skyrmions can be unambiguously determined. A rigorous theoretical treatment demonstrates the suitability for studying all types of skyrmion-hosting materials, and other topological structures.

The crowning Chap. 6 introduces the Dichroism Extinction Rule (DER), relating the circular dichroism in resonant magnetic scattering with the structure of the motif. The method, as a dichroic effect, is sensitive to all types of modulated spin spirals in magnetic materials. These spirals are ubiquitous in almost all areas of condensed matter physics, reaching from helimagnets, multiferroics, and molecular magnets to frustrated systems, and indeed also comprise skyrmions, which can be thought of as being composed of chiral bases. Magnetic microscopy methods, such as Lorentz transmission electron microscopy, scanning probe microscopy, Kerr microscopy, photoemission electron microscopy, and x-ray microscopy failed to retrieve the full information of such structures, mostly due to their limited spatial resolution, the accessibility to all three components of the magnetisation vector and/or the small probing area, unsuited for the length scales of interest. On the other hand, neutron and x-ray diffraction techniques, which are commonly used, cannot directly reveal the underlying type of spin spiral, as data refinement, computational modelling, as well as theoretical comparisons have to be performed, yielding no unique answer. In stark contrast to these disadvantages, Shilei's novel light-matter interaction-based DER principle offers a one-to-one correspondence between spin structure and measured signal, thus allowing for the real-space spin structure to be unambiguously determined.

An important focus of this book is on experimental x-ray scattering techniques that have been used to study skyrmions across all length scales. The introduction of linear and circular polarisation give rise to new REXS-based techniques that elegantly relate the geometrical properties of circular dichroism to the underlying microscopic spin structure of a material. Shilei's discoveries will have a profound impact on skyrmion science and beyond, as the widely applicable experimental techniques pave the way for the in-depth study of topological magnetic structures in general.

Oxford, UK   Prof. Thorsten Hesjedal
April 2018

Abstract

This work focuses on characterising the chiral and topological nature of magnetic skyrmions in non-centrosymmetric helimagnets. In these materials, the skyrmion lattice phase appears as a long-range-ordered, close-packed lattice of nearly millimetre-level correlation length, while the size of a single skyrmion is 3–100 nm. This is a very challenging range of length scales (spanning 5 orders of magnitude from tens of nm to mm) for magnetic characterisation techniques. As a result, only three methods have been proven to be applicable for characterising certain aspects of the magnetic information: neutron diffraction, electron microscopy, and magnetic force microscopy. Nevertheless, none of them reveals the complete information about this fascinating magnetically ordered state. On the largest scale, the skyrmions form a three-dimensional lattice. The lateral structure and the depth profile are of importance for understanding the system. On the mesoscopic scale, the rigid skyrmion lattice can break up into domains, with the domain size about tens to hundreds of micrometres. The information of the domain shape, distribution, and the domain boundary is of great importance for a magnetic system. On the smallest scale, a single skyrmion has an extremely fine structure that is described by the topological winding number, helicity angle, and polarity. These pieces of information reveal the underlying physics of the system, and are currently the focus of spintronics applications. However, so far, there is no experimental technique that allows one to quantitatively study these fine structures. It has to be emphasised that the word 'quantitative' here means that no speculations have to be made and no theoretical modelling is required to assist the data interpretation—what has been measured must be straightforward, and give a unique and unambiguous answer.

Motivated by these questions, we developed soft x-ray scattering techniques that allow us to acquire much deeper microscopic information of the magnetic skyrmions—reaching far beyond what has been possible so far. We will show that by using only one technique, all the information about the magnetic structure (spanning 5 orders of magnitude in length) can be accurately measured. The key development of the thesis is the *Dichroism Extinction Rule*, which is summarised in Chap. 6, and quintessentially summarises this work.

Parts of this thesis have been published in the following journal articles:

1. Chapter 1: S. Zhang, A. A. Baker, S. Komineas, and T. Hesjedal, "Topological computation based on direct magnetic logic communication", Sci. Rep. **5**, 1773 (2015).
2. Chapter 2: A. I. Figueroa, S. L. Zhang, A. A. Baker, R. Chalasani, A. Kohn, S. C. Speller, D. Gianolio, C. Pfleiderer, G. van der Laan, and T. Hesjedal, "Strain in epitaxial MnSi films on Si(111) in the thick film limit studied by polarization-dependent extended x-ray absorption fine structure", Phys. Rev. B **94**, 174107 (2016).
3. Chapter 3: S. L. Zhang, A. Bauer, H. Berger, C. Pfleiderer, G. van der Laan, and T. Hesjedal, "Imaging and manipulation of skyrmion lattice domains in Cu_2OSeO_3", Appl. Phys. Lett. **109**, 192406 (2016).
4. Chapter 2: S. L. Zhang, A. Bauer, H. Berger, C. Pfleiderer, G. van der Laan, and T. Hesjedal, "Resonant elastic x-ray scattering from the skyrmion lattice in Cu_2OSeO_3", Phys. Rev. B **93**, 214420 (2016).
5. Chapter 3: S. L. Zhang, A. Bauer, D. M. Burn, P. Milde, E. Neuber, L. M. Eng, H. Berger, C. Pfleiderer, G. van der Laan, and T. Hesjedal, "Multidomain Skyrmion Lattice State in Cu_2OSeO_3", Nano Lett. **16**, 3285 (2016).
6. Chapter 2: T. Lancaster, F. Xiao, Z. Salman, I. O. Thomas, S. J. Blundell, F. L. Pratt, S. J. Clark, T. Prokscha, A. Suter, S. L. Zhang, A. A. Baker, and T. Hesjedal, "Transverse field muon-spin rotation measurement of the topological anomaly in a thin film of MnSi", Phys. Rev. B **93**, 140412 (2016).
7. Chapter 2: S. L. Zhang, R. Chalasani, A. A. Baker, N.-J. Steinke, A. I. Figueroa, A. Kohn, G. van der Laan, and T. Hesjedal, "Engineering helimagnetism in MnSi thin films", AIP Adv. **6**, 015217 (2016).
8. Chapter 5: S. L. Zhang, G. van der Laan, and T. Hesjedal, "Direct experimental determination of spiral spin structures via the dichroism extinction effect in resonant elastic soft x-ray scattering", Phys. Rev. B **96**, 094401 (2017).
9. Chapter 2: S. L. Zhang, I. Stasinopoulos, T. Lancaster, F. Xiao, A. Bauer, F. Rucker, A. A. Baker, A. I. Figueroa, Z. Salman, F. L. Pratt, S. J. Blundell, T. Prokscha, A. Suter, J. Waizner, M. Garst, D. Grundler, G. van der Laan, C. Pfleiderer, and T. Hesjedal, "Room-temperature helimagnetism in FeGe thin films", Sci. Rep. **7**, 123 (2017).
10. Chapter 4: S. L. Zhang, G. van der Laan, and T. Hesjedal, "Direct experimental determination of the topological winding number of skyrmions in Cu_2OSeO_3", Nat. Commun. **8**, 14619 (2017).
11. Chapter 5: S. L. Zhang, G. van der Laan, W. W. Wang, A. A. Haghighirad, and T. Hesjedal, "Direct observation of twisted surface skyrmions in bulk crystals", Phys. Rev. Lett. **120**, 227202 (2018).
12. Chapter 5: S. L. Zhang, G. van der Laan, J. Mueller, L. Heinen, M. Garst, A. Bauer, H. Berger, C. Pfleiderer, and T. Hesjedal, "Reciprocal space tomography of 3D skyrmion lattice order in a chiral magnet", Proc. Natl. Acad. Sci. U.S.A. **115**, 6386 (2018).

Acknowledgements

The aesthetic standards of physicists have been changing over time. I entered the field of condensed matter physics in an era when materials are dominating the experimental physics, and new materials symbolise the cutting-edge science. This strong conceptual wave had been driving me constantly looking for new materials in the early years of my D.Phil. study. I feel very fortunate to be educated under a thin film growth background, in which I have the complete freedom to choose what to synthesise, and to only study the materials that I am personally interested. This is largely due to the kind support and encouragement from my D.Phil. supervisor (Thorsten Hesjedal)—and I have been always appreciating that. This extreme freedom of choice is the major reason I decided to stay in the field as my career.

The other thing that I am quite pleased about my D.Phil. work is the project I chose (of course the magnetic skyrmions), which turns out to be a very satisfactory decision. I have to admit that the only reason I initially chose this project was merely because it looked extremely beautiful. Nevertheless, the major problem of this project was not from the materials but from the experimental techniques. The accessible experimental techniques used to study the skyrmions all have their own limitations. Most importantly, I could not get my hands on them easily, even I had many skyrmion-carrying materials grown or collected. This was the time when I diverted my efforts from materials growths to the developments of new experimental techniques. For this project, I thought it would make more sense to characterise the materials properly using new methods, than having lots of new materials but with ambiguous results. This was why I got addicted to x-rays—the light that illuminates the skyrmions. We first saw the skyrmion lattice diffraction pattern in MnSi in 2013, using a very strange geometry. After we saw the same thing on Cu_2OSeO_3, I had full confidence with this technique. This also made us feel cool, as using x-rays (not neutrons or electrons) to characterise the skyrmions was very new at that time (and is even true for now).

However, at the same time, I was aware of the extra degree of freedom of x-rays: its polarisation. Then I started to explore more possibilities from this perspective. Thankfully this is what the people in Clarendon are good at. The casual and random discussions with them are always helpful to design a suitable method for my own system. The other thing I have to confess is that I learned most of my scattering knowledge from an x-ray spectroscopist, and I feel lucky about it. First, he is one of the greatest x-ray magnetic spectroscopists (Gerrit van der Laan, of course); second, learning the magnetic scattering from a spectroscopic point of view is extremely useful for understanding many underlying physics of the entire process; third, due to his great impact on magnetic dichroism, I became a huge fan of circular dichroism phenomena. This is why I tried to apply dichroism in a resonant magnetic diffraction experiment, and subsequently developed the dichroism extinction rule. While the theory is written up within a few pages, the experimental realisation demands massive efforts and dedications. Luckily my supervisor is the greatest experimentalist for me, who has been always giving me important advices and help to my project. The set-ups and trials are painful, however we managed to measure those delicate microscopic information with high accuracy eventually, and the theoretical rule was well-applied. Until then, the skyrmions can be characterised in great details (at least to the level that I am happy with) purely by x-rays, without any other complementary methods. Moreover, up to now, no other techniques can measure these fine information. This is the point that I am comfortable to write the story.

I would like to deeply thank for the kind help from the following persons, without whom this work would not be possibly made: Gerrit van der Laan, whose wisdom on x-ray magnetic circular dichroism, as well as many educations and discussions, were great inspirations for my thesis. Christian Pfleiderer, Peter Böni, Andreas Bauer, Birgit Wiedemann, Ioannis Stasinopoulos and Felix Rucker from Technical University of Munich, for providing high-quality single crystals and many other collaborative works. Achim Rosch, Jan Müller from University of Cologne for the beautiful theoretical works. Paul Steadman, Steve Collins, Chris Nicklin, Adriana Figueroa, Raymond Fan, David Burn, Peter Bencok, Mark Sussmuth, Roberto Boada-Romero from Diamond Light Source; and Sean Langridge, Nina-Juliane Steinke from ISIS neutron source, for the kind beamline support and large amount of beamtime efforts. Hans Fangohr, Weiwei Wang and Marijan Beg from University of Southampton for the kind supports of the numerical simulations. Paolo Radaelli, Yulin Chen and Radu Coldea from Clarendon Laboratory, University of Oxford, for fruitful discussions and various collaborative works. Also the kind colleagues in Clarendon, Amir Abbas Haghighirad, Alexander Baker, Francis Chmiel, Junjie Liu, Jian Huang, Liam Collins-McIntyre, Liam Duffy, Marein Rahn, Piet Schönherr, Roger Johnson and Tsz Cheong Fung for various help on my thesis and the stimulating daily academic environment. Moreover, I would like to thank Stavros Komineas (University of Crete) for bringing me into the world of micromagnetism, Helmuth Berger (EPFL) for providing high-quality Cu_2OSeO_3 crystals, Tom Lancaster (University of Durham) for the muon work and many valuable discussions, Hans-Benjamin Braun (University College Dublin) for the help on the analytical solution of the x-ray skyrmion scattering form factor, Amit Kohn and Rajesh Chalasani (Tel Aviv University) for the transmission electron microscopy work and Susannah Speller (University of Oxford) for the electron diffraction work.

Lastly, and most importantly, I would like to thank my mentor Thorsten Hesjedal, who has been constantly inspiring me, helping me and looking after me not only in academics but also many aspects in life, which are precious treasures for my career. We truly had productive and fun time in Oxford.

Oxford, UK   Shilei Zhang
October, 2016

Contents

Acronyms

ACD	Azimuthal circular dichroism
AHE	Anomalous Hall effect
CD	Circular dichroism
CRP	Common rotation plane
DER	Dichroism extinction rule
DMI	Dzyaloshinskii-Moriya interaction
EBSD	Electron backscatter diffraction
FMR	Ferromagnetic resonance
LLG	Landau–Lifshitz–Gilbert
LLGS	Slonczcwski Landau–Lifshitz Gilbert
LTEM	Lorentz transmission electron microscopy
MFM	Magnetic force microscopy
OHE	Ordinary Hall effect
PAM	Polarisation-azimuthal-map
PCD	Polar circular dichroism
RASOR	Reflectivity and advanced scattering from ordered regimes
REXS	Resonant elastic x-ray scattering
RSM	Reciprocal space mapping
r.l.u.	Reciprocal lattice unit
SANS	Small angle neutron scattering
SLLG	Stochastic Landau–Lifshitz–Gilbert
SOC	Spin-orbit coupling
TDP	Topology determination principle
THE	Topological Hall effect

Chapter 1
The Story So Far

Over the past years, an increasing number of materials have been identified to host magnetic skyrmions. Before starting, it is therefore necessary to properly classify the magnetic skyrmions found so far, and sort them into the categories using an experimentalist's point of view.

In general, a magnetic skyrmion specifies a $N \geq 1$ topological winding number entity that is composed of a magnetisation vector field in a classical model [1]; or in a quantum-mechanics framework, the entity is composed of spins [2, 3]. Therefore, any material that carries such a magnetic phase can be categorised using this terminology, regardless of how distorted the topological object appears, or how disordered the assembly of multiple skyrmions appear, or how extreme the conditions the material requires to reach such a phase. Consequently, there are two major classes of magnetic skyrmion systems: (I) systems with broken inversion symmetry, in which the Dzyaloshinskii-Moriya interaction (DMI) is present, called DMI-skyrmions, and (II) systems that are inversion symmetric, where the DMI is not present, usually called bubble skyrmions.

In the DMI-skyrmion class, there are three types of materials systems: (1) In certain single crystals, the three-dimensional crystalline environment supports the DMI [4–6]; (2) Surface-DMI-skyrmion materials, where the inversion symmetry is naturally broken at the surface of a quasi-two-dimensional magnetic thin film system [7]; (3) Interfacial-DMI-skyrmion materials, in which the inversion symmetry is broken at the very boundary between two different crystalline phases, which may both be centrosymmetric [8–13].

In the bubble-skyrmions class, there are also several species: (1) Polar-skyrmions, which are formed in many single-crystalline bulk polar magnets [14–17]; (2) Classical magnetic bubble thin films, which were a popular topic in the 1970s [18]; (3) Artificial bubble skyrmions, in which the skyrmions are designed objects, using material engineering, such as heterostructures or nano-patterning [19–22]. Note that no further subcategories are introduced, though one could further divide the species into the different material families.

© Springer Nature Switzerland AG 2018
S. Zhang, *Chiral and Topological Nature of Magnetic Skyrmions*,
Springer Theses, https://doi.org/10.1007/978-3-319-98252-6_1

In this work we solely concentrate on the materials of the (I)–(1) type, more specifically, the $P2_13$ crystallographic family (otherwise called noncentrosymmetric helimagnets). Therefore, in the context of this thesis, 'skyrmions' mean bulk-DMI-skyrmions, unless otherwise specified. The magnetism in this family of materials shares many similarities, and can be described by a universal theory [4–6].

Several excellent review articles and books provide an introduction to this field, including: '*Topological Properties and Dynamics of Magnetic Skyrmions*' by Nagaosa and Tokura [2]; '*Topological Skyrmion Dynamics in Chiral Magnets*' by Garst [5]; '*Generic Aspects of Skyrmion Lattices in Chiral Magnets*' by Bauer and Pfleiderer [4]; and '*Skyrmions in Magnetic Materials*' by Seki and Mochizuki [6]. In the following we will summarise the well-established theoretical treatments that are needed to describe the magnetic orders and magnetic properties in the $P2_13$ system.

1.1   Theoretical Treatments

1.1.1   Skyrme Field in the Non-linear Sigma Model

The original form of the skyrmion solution was described by Tony Skyrme (1922–1987) [23, 24], and was used to deal with the nucleons, indeed very far away from magnetism. Nevertheless, we would like to spend a few words introducing this theory, and one finds many similarities when applying the same concept to a continuum spin system. The simplest Skyrme field $\varphi(\mathbf{x})$ is a scalar field that takes $SU(2)$ symmetry, constructed based on the non-linear sigma model, where $\mathbf{x} = (\mathbf{r}, t)$ is a four-vector [25]. It was intended to describe the nucleons (protons and neutrons) using a soliton solution that is composed of three nearly massless $SO(3)$-symmetric pion fields $\pi_1(\mathbf{x}), \pi_2(\mathbf{x})$ and $\pi_3(\mathbf{x})$. Using an extra field $\sigma(\mathbf{x})$ that satisfies $\sigma^2 + \pi_1^2 + \pi_2^2 + \pi_3^2 = 1$, the $SO(3)$ symmetry is parametrised by a three-sphere. The Skyrme field can then be written as [26, 27]:

$$\varphi = \begin{pmatrix} \sigma + i\pi_3 & i\pi_1 + \pi_2 \\ i\pi_1 - \pi_2 & \sigma - i\pi_3 \end{pmatrix}. \tag{1.1}$$

This is in essence the classical trick that makes up a $SU(2)$ group from its homomorphic $SO(3)$ group using Pauli matrices [1]. Also, Lorentz invariance in $3+1$ dimensions will hold. However, only the static solution is important here, so we can ignore the time-dependence. The effective current of this field then takes the form of $R_\mu = (\partial_\mu \varphi)\varphi^{-1}$. The Lagrangian density under this model is written as:

$$\mathscr{L} = -\frac{1}{2}\mathrm{Tr}(R_\mu R^\mu) + \frac{1}{16}\mathrm{Tr}([R_\mu, R_\nu][R^\mu, R^\nu]) + m^2\mathrm{Tr}(\varphi - 1), \tag{1.2}$$

where m is a dimensionless pion mass parameter, and the boundary condition requires $\varphi(|\mathbf{r}| \to \infty) \to 1$. Therefore, the Lagrangian only contains three terms: a kinetic term that is quadratic in field derivatives; the Skyrme term that is quartic in the field derivatives; and a field potential energy term that is explicit in the pion mass. The first two terms preserve the chiral symmetry, and it is the pion mass term that breaks the chiral symmetry of the system.

Consequently, the energy density w of a time-independent skyrmion field can be written as:

$$w = -\frac{1}{2}\text{Tr}(R_i R_i) - \frac{1}{16}\text{Tr}([R_i, R_j][R_i, R_j]) - m^2\text{Tr}(\varphi - 1). \quad (1.3)$$

This expression preserves both translational and rotational symmetry. Moreover, an important quantity, the baryon number N that takes integer numbers, guarantees that a soliton solution exists. In other words, the field solutions for nucleons and nuclei are 'protected' by the baryon number in order for them to behave like particles. The Baryon number is written as:

$$N = -\frac{1}{24\pi^2} \int \varepsilon_{ijk}\text{Tr}(R_i R_j R_k)d^3\mathbf{r}, \quad (1.4)$$

where ε_{ijk} is the antisymmetric tensor, and the integration is carried out over real space. The boundary condition of $\varphi(|\mathbf{r}| \to \infty) \to 1$ holds. Equation (1.4) essentially counts how many times real-space has completely covered the parameter space φ. It is then a topological quantity, also called topological charge. Here, the parameter space is a scalar, which is parametrised by a three-sphere, while real space is three-dimensional. It alters Eq. (1.3) such that $E \geq 12\pi^2|N|$.

In case of the field configuration, an axial-symmetric ansatz is used for the system. Therefore, we use polar coordinates $\mathbf{r} = (\rho, \theta, \psi)$, and $z = \tan\frac{\theta}{2}\exp(i\psi)$. Then the ansatz of the field configuration has a solution of [25–27]:

$$\varphi(\rho, z) = \begin{pmatrix} \cos f(\rho) + i\sin f(\rho)(\frac{1-|R|^2}{1+|R|^2}) & i\sin f(\rho)(\frac{2|R|}{1+|R|^2}) \\ -i\sin f(\rho)(\frac{2|R|}{1+|R|^2}) & \cos f(\rho) - i\sin f(\rho)(\frac{1-|R|^2}{1+|R|^2}) \end{pmatrix}, \quad (1.5)$$

where $f(\rho)$ is the radial profile function describing the radial property of φ. It satisfies the boundary conditions of $f(0) = \pi$, and $f(\infty) = 0$. $R(z)$ is the angular function that accounts for the 'winding' of the field. The solution can be obtained numerically by minimising the energy in Eq. (1.3) for a specific baryon number. For example, $R(z) = z^2$ gives the $N = 1$ solution, which is a spherically symmetric, hedgehog-like solution, as shown in Fig. 1.1a. The angular part $R(z)$ for higher N's, obtained by minimising the energy, can be found in Table 1.1.

Such solutions $\varphi(\mathbf{r})$ are called *skyrmions*. Looking at Fig. 1.1, one can see that the topological quantity N plays an important role for making sure the field solution has particle properties; and for governing the underlying symmetry of the field. The reason why N is closely related to topology can be directly seen from the shapes of the baryon number density (also called vorticity) distributions. For $N = 2$ skyrmions, one has to 'punch a hole' into a $SO(3)$ ball, in order to get a $SO(2)$ doughnut. The

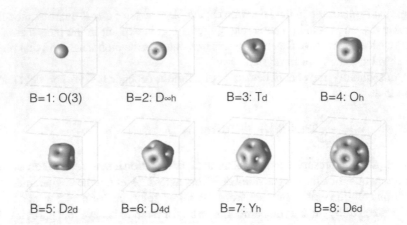

B=1: O(3)   B=2: D∞h   B=3: Td   B=4: Oh

B=5: D2d   B=6: D4d   B=7: Yh   B=8: D6d

Fig. 1.1 Skyrmion solutions of non-linear sigma model specified by Eq. (1.2). The Skyrme field φ is classified by the baryon number N. $1 \leq N \leq 8$ is plotted with the isosurfaces of the baryon number density $-\varepsilon_{ijk}\mathrm{Tr}(R_i R_j R_k)/(24\pi^2)$, with different symmetric properties. Reprinted with permission from Ref. [28]. Copyright 2018 by The Authors

Table 1.1 Energy minimising angular function $R(z)$ for different baryon numbers. The constants a and b are obtained numerically. From [26]

N	$R(z)$	Symmetry
1	z	$O(3)$
2	z^2	$O(2)$
3	$\frac{i\sqrt{3}z^2-1}{z(z^2-\sqrt{3})}$	T_d
4	$\frac{z^4+i2\sqrt{3}z^2+1}{z^4-i2\sqrt{3}z^2+1}$	O_h
5	$\frac{z(z^4+bz^2+a)}{az^4-bz^2+1}$	D_{2d}
6	$\frac{z^4+a}{z^2(az^4+1)}$	D_{4d}
7	$\frac{z^5-a}{z^2(az^5+1)}$	Y_h
8	$\frac{z^6-a}{z^2(az^6+1)}$	D_{6d}
.	.	.
.	.	.

'punching hole' operation represents an extremely violent process, and indicates a high energy barrier between $N = 1$ and $N = 2$ skyrmions: one cannot transform them into each other by a gentle, smooth, continuous local surgery. Analogously, $(N - 1)$ holes have to be punched for N-skyrmions.

The points we want to make from this seemingly irrelevant theory are as follows: (1) the non-linear sigma model is a general effective field theory that captures the spontaneous symmetry breaking of an $O(n)$ system. Therefore, it also accounts for a low-energy, long-wavelength microscopic model in condensed matter physics, if the system has a field nature. (2) A skyrmion solution is a static, axial-symmetric field configuration that satisfies both stability and energetic preference. The stability is achieved by having a certain topological winding protection, while the energy preference is achieved by adjusting the radial profile and the angular function. Depending on the effective potential, it can be non-chiral (massless pion), or chiral, in order to minimise the energy. (3) All the possible topological configurations form a group, called homotopy group, distinguished by the winding number. The winding number is mainly encoded in the angular functions.

As the thesis title suggests, we will be focusing on two remarkable properties of this fascinating field configuration: chirality and topology. Next, we will use a similar treatment to write down the skyrmion field in magnetism, with the goal to measure these two features by experiments.

1.1.2 The Theory of Magnetic Skyrmions

The theoretical treatment of the Skyrme model can be simplified down to a few steps. First, one has to identify the system as an effective field, with an order parameter $\varphi(\mathbf{r})$, and only consider the static solutions. Second, the Lagrangian is written down, from which a time-independent energy functional is obtained. Third, taking an axial-symmetric ansatz, the boundary conditions are defined. Then consider the topological stability by making up an angular function. Last, find the radial profile, as well as the angular function that minimises the energy.

We now apply this approach to the magnetism of $P2_13$ helimagnets. The magnetisation can come from itinerant electrons (in metals) [29–47] or local spins (in insulators) [48–53]. However, the weak spin-orbit coupling (SOC) indicates that an effective continuum field can be used to describe the magnetic moment density [2, 5, 6]. The order parameter is therefore the vector magnetisation $\mathbf{m}(\mathbf{r}, t) = [m_1(\mathbf{r}, t), m_2(\mathbf{r}, t), m_3(\mathbf{r}, t)]$, where its amplitude, the saturation magnetisation M_S, keeps as a constant, i.e., $m_1^2 + m_2^2 + m_3^2 = M_S^2$. Further, the broken inversion symmetry induces the reduced Dzyaloshinskii-Moriya interaction (DMI) D [54, 55]. The sign of D arises from the left- or right-hand chirality of the crystallographic system, and defines the chirality of the magnetism [2]. The amplitude of D is related to the strength of the SOC. Due to the weak SOC in $3d$ $P2_13$ compounds, the competition between the Heisenberg-type exchange interaction J and the DMI results in a spatially modulated magnetisation configuration with a wavevector $q_h \sim D/J$ [56]. Therefore, in the non-linear sigma model, the Lagrangian density of this field can be written as [5, 56]:

$$\mathcal{L} = \frac{J}{2}(\nabla \mathbf{m})^2 + D\mathbf{m}(\nabla \times \mathbf{m}) - \mathbf{m} \cdot \mathbf{B}, \tag{1.6}$$

where $\mathbf{B}$ is the external magnetic field. The first term describes the exchange interaction, the second term describes the antisymmetric DMI, and the third term is the

Zeeman term. Consequently, the static energy density with $\mathbf{m}(\mathbf{r})$ can be obtained as: $\frac{J}{2}(\nabla\mathbf{m})^2 + D\mathbf{m}(\nabla \times \mathbf{m}) - \mathbf{m} \cdot \mathbf{B}$. This is the core energy term that we need to minimise. Moreover, if we also taking the phase transition into account, Ginzburg-Landau theory is called for [34]. The complete form of the free energy is the written as:

$$ w = \frac{J}{2}(\nabla\mathbf{m})^2 + D\mathbf{m}(\nabla \times \mathbf{m}) - \mathbf{m} \cdot \mathbf{B} + \frac{r}{2}\mathbf{m}^2 + \frac{u}{4!}\mathbf{m}^4 + w_{\text{cubic}}, \tag{1.7} $$

$$ E = \int w \, d^3\mathbf{r} + E_{\text{dipolar}}, \tag{1.8} $$

where r and u are constants, w_{cubic} is the magnetic anisotropy arising from the cubic crystalline environment that breaks the rotation symmetry, and E_{dipolar} is the energy accounting for the non-local dipole-dipole interaction. While the core energy term is given by Eq. (1.6), additional energy terms may play important roles in different scenarios. For example, if the system is close to the phase transition, the $\mathbf{m}^2$ and $\mathbf{m}^4$ terms become important [34, 57–59], and if a three-dimensional system is considered, w_{cubic} becomes important [60]. On the other hand, the dynamics of the system, described in the micromagnetic theoretical framework, is best captured by considering E_{dipolar} [61].

Let us start from the 'thin film' limit, meaning that the real space is two-dimensional. In this limit, the anisotropy term takes the two-dimensional form [36, 62–64]. Furthermore, let us assume that the system is well below T_c, meaning that no phase transition is to be expected. Therefore, the $\mathbf{m}^2$ and $\mathbf{m}^4$ terms can be ignored. This reduces the total Ginzburg-Landau free energy to Eq. (1.6). We can now start to construct a non-dispersive static soliton ansatz solution for Eq. (1.6), meaning that a certain form of topology of the field distribution is to be imposed, namely a two-dimensional real space, with the order-parameter-space being a two-sphere (the vector magnetisation has fixed length. Therefore, all the possible values of the magnetisation form a sphere) [1, 65]. This makes the situation simpler than for the baryon model, as the description of the topology here is essentially only counting how many times a two-dimensional real-space has fully 'wrapped' around a sphere. If we write up an angular quantity [66–70]

$$ \mathcal{V} = (\partial_j\mathbf{m} \times \partial_i\mathbf{m}) \cdot \mathbf{m}, \tag{1.9} $$

we have obtained the topological number density, which is the counterpart to the baryon number density, accounting for the solid angle that is spanned by neighbouring spin vectors. By integrating the total spanned solid angle over real space, the 'wrapping' problem is solved. The full coverage of the order-parameter-space sphere has a value of 4π. Therefore, the topological winding number for the three-vector field in two-dimensional real space can be written as:

$$ N = \frac{1}{4\pi} \int \mathcal{V} \, d^2\mathbf{r} = \frac{1}{4\pi} \int (\partial\mathbf{m}_j \times \partial_i\mathbf{m}) \cdot \mathbf{m} \, d^2\mathbf{r}. \tag{1.10} $$

All the possible integers N for this setup form the homotopy group $\pi_2(S^2)$ [1, 65]. It therefore specifies how difficult it is to transform the vector field distributions between the solutions with distinct N.

Next, we consider the exact field configuration $\mathbf{m}(\mathbf{r})$. We expect the solution to be axial-symmetric. Therefore, polar coordinates are used for the real space (ρ, ψ). To satisfy the topological argument above, the magnetisation components have to be written in the form of [2, 62, 70, 71]:

$$
\begin{aligned}
m_1(\rho, \psi) &= M_S \sin[\theta(\rho)] \cos[N(\psi + \chi)], \\
m_2(\rho, \psi) &= M_S \sin[\theta(\rho)] \sin[N(\psi + \chi)], \\
m_3(\rho, \psi) &= M_S \lambda \cos[\theta(\rho)],
\end{aligned}
\tag{1.11}
$$

where M_S is the saturation magnetisation, and $\theta(\rho)$ is the radial profile of the field, which especially describes the radial function of the m_3 component. The boundary conditions take the form of $\theta(0) = \pi$, and $\theta(\infty) = 0$. The axial-symmetric property imposes that the $\theta(\rho)$ function satisfies the Euler-Lagrange equation [62, 72]:

$$
\frac{J}{2} \left(\frac{d^2\theta}{d\rho^2} + \frac{1}{\rho} \frac{d\theta}{d\rho} - \frac{1}{\rho^2} \sin\theta \cos\theta \right) - \frac{B}{2M_S} \sin\theta + \frac{D}{\rho} \sin^2\theta = 0 .
\tag{1.12}
$$

The angular properties are expressed by the $\cos[N(\psi + \chi)]$ and $\sin[N(\psi + \chi)]$ terms, in which N is the topological winding number. χ is defined as helicity angle, which describes the 'phase shift' of the winding of the in-plane components. $\lambda = \pm 1$ is termed the polarity, and it specifies whether the m_3 component in the centre of the field is positive or negative.

Let us first prove that this form of a solution can indeed carry non-zero topological charge, by plugging Eq. (1.11) into Eq. (1.10) [2]:

$$
\begin{aligned}
&\frac{1}{4\pi} \int (\partial \mathbf{m}_j \times \partial_i \mathbf{m}) \cdot \mathbf{m} \, d^2\mathbf{r} \\
=& \frac{1}{4\pi} \int_0^\infty d\rho \frac{d\theta(\rho)}{d\rho} \sin\theta(\rho) \int_0^{2\pi} d\psi \frac{d[N(\psi + \chi)]}{d\psi} \\
=& N .
\end{aligned}
\tag{1.13}
$$

Therefore, the ansatz in Eq. (1.11) indeed describes a 'topologically protected' soliton solution with integer winding number N. Field configurations with different winding numbers are illustrated in Fig. 1.2, where it is obvious that the topological nature is represented by the number of 'buckling-like kinks'. For example, a $N = 2$ magnetic skyrmion has one 'buckling-like kink' starting from the centre, dividing the field up through the right-hand boundary. Similarly, $(N - 1)$ 'kinks' are imposed for N skyrmion. In fact, this reflects the periodicity of the angular function, which is governed by N.

Next, we minimise the energy by choosing an appropriate radial profile function and proper parameters for the angular concentrating on the $N = 1$ skyrmion. There

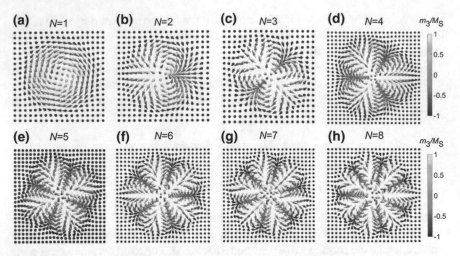

Fig. 1.2 Magnetic skyrmion solutions specified by Eq. (1.11). The field solution $\mathbf{m}(\mathbf{r})$ is classified by the winding number N. The field configurations for $1 \leq N \leq 8$ are plotted

are various radial profiles one can chose (either from analytical approximation results or numerical results), which lead to different total energies [5, 62, 68, 69, 71, 73]. Therefore, whether the formation of the skyrmion texture is an energetically preferred state or whether it is an excitation state is determined by the choice of the $\theta(\rho)$ function. Depending on the particular regime whether the skyrmion is stable of not, the system can accommodate many skyrmions that fill out the 'ferromagnetic' background, forming a lattice state [63, 73], as shown in Fig. 1.3a. In this scenario, only $\chi = \pm\pi/2$ minimises the energy [2, 5, 73]. The sign of χ is governed by the combination of the sign of λ and the sign of D, i.e., the chirality that is inherited from the crystalline symmetry. Consequently, the skyrmion becomes chiral, as shown in Fig. 1.2a. In the two-dimensional case where the thin film is perpendicular to the magnetic field, the skyrmion lattice is favoured at zero temperature and small fields. The core magnetisation of the skyrmion is always antiparallel to the magnetic field, as shown in Fig. 1.3a and b.

In summary, the skyrmion solution in the magnetism framework resembles the original baryon field skyrmion to a large degree. First, the non-linear sigma model captures the main features of the system, i.e., a stable soliton solution accounts for the spontaneous symmetry breaking out of a $SO(2)$ (or $SO(3)$ for the bulk case) system. Second, a skyrmion is an axial-symmetric field which is a stable state. It can be non-chiral, or chiral, depending on the energy configuration of the system. In the weak SOC limit in which the DMI induces a long-wavelength modulated state, $N = 1$, $\chi = \pm 90°$ chiral skyrmions are present. The radial profile adjusts the total energy, and the angular function encodes the topological winding number. Third, the homotopy group for such a system (two-dimensional real space, 3-sphere order parameter space) is $\pi_2(S^2)$.

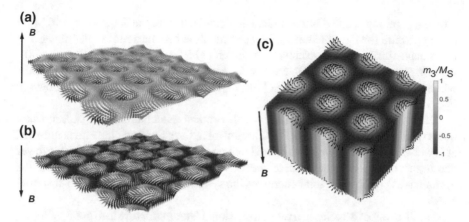

Fig. 1.3 Skyrmion crystal configuration. **a, b** $N = 1$ skyrmions form periodic lattice, filling up the two-dimensional space that is composed of the ferromagnetic background. The polarity of the skyrmions depend on the polarity of the external magnetic field, i.e., the core of the skyrmion is always antiparallel to the field. **c** Three dimensional skyrmion crystal that is constructed based on repeatedly stacking the two-dimensional skyrmion lattice into a third dimension, showing the 'tube-like' structure

While the main difference between baryonic and magnetic skyrmions may come from the scalar-versus-vector nature of the order parameter, one can always use a complex scalar field of $\varphi(\mathbf{r})$ that accounts for the stereographic projection to describe the magnetic skyrmion field equivalently [68, 70]. In this representation,

$$\varphi = \frac{m_1 + im_2}{1 + m_3} = f(\rho)e^{iN(\psi + \chi)} ,$$
$$f(\rho) = \frac{\sin \theta(\rho)}{1 + \lambda \cos \theta(\rho)} , \tag{1.14}$$

where $f(\rho)$ is the radial profile, and therefore Eq. (1.14) shares a similar form with Eq. (1.5). The major difference comes from the dimensionality.

1.1.3 Magnetism of P2₁3 Helimagnets

In a real material (for both bulk and thin film form), the system should be three-dimensional. In this case, a conical spin helix state with modulation wavevector $\mathbf{q}_h$ along the field direction is more stable than the skyrmion solution, written as [34, 72]

$$m_1(\mathbf{r}) = M_S \cos \xi ,$$
$$m_2(\mathbf{r}) = M_S \sin \xi \cos(\mathbf{q}_h \cdot \mathbf{r}) , \tag{1.15}$$
$$m_3(\mathbf{r}) = M_S \sin \xi \sin(\mathbf{q}_h \cdot \mathbf{r}) ,$$

where $\mathbf{q}_h$ is along $\mathbf{x}$, and ξ is the conical angle, which relates to $\cos \xi \sim B/(M_S)q_h D$ [5], i.e., while the field increases, the conical angle becomes more tilted towards a ferromagnetically aligned structure, as shown in Fig. 1.4b. Therefore, $0 \leq \xi \leq 90°$. At zero magnetic field ($B = 0$), the system prefers $\xi = 90°$. This is the helical state, which forms the proper-screw-type [74, 75] modulated spin structure, as shown in Fig. 1.4a.

This is indeed consistent both with theoretical calculations and experimental observations [2, 4–6]. In a bulk $P2_1 3$ helimagnet, the skyrmion phase does not exist at lower temperatures, while the helical and conical states are more stable (for lower and higher fields). Only at temperatures close to T_c, the strong fluctuations from the paramagnons modify the total energy of the system, leading to a skyrmion-favoured state.

Another way of describing a two-dimensional lattice structure is to use a multiple-$\mathbf{q}$ spin-density-wave-like model [34]. This is analogous to the interference phenomena of several cross-propagating waves that produce standing wave-like vortex structures. It can be written as:

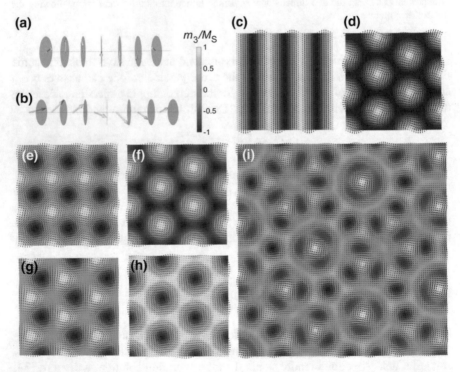

Fig. 1.4 one-dimensional modulated structure and the multiple-$\mathbf{q}$ structure. **a** one-dimensional helical structure. **b** one-dimensional conical structure. **c** one-dimensional helical lattice. **d** two-dimensional skyrmion lattice generated by Eq. (1.11). **e** 4-$\mathbf{q}$ structure. **f** Triple-$\mathbf{q}$, or 6-$\mathbf{q}$ structure. **g** Triple-$\mathbf{q}$ structure with different helical phases. **h** Triple-$\mathbf{q}$ structure with π phase difference compared with (**f**). **i** 5-$\mathbf{q}$ structure, forming a magnetic quasi-crystal

$$\mathbf{m}(\mathbf{r}) = \frac{M_S}{n} \sum_{i=1}^{n} [\mathbf{n}_{i1} \cos(\mathbf{q}_{hi} \cdot \mathbf{r} + \kappa_i) + \mathbf{n}_{i2} \sin(\mathbf{q}_{hi} \cdot \mathbf{r} + \kappa_i)] + \mathbf{m}_{\text{net}}, \quad (1.16)$$

where the total number of the propagating density-waves is n. The unit vectors $\mathbf{n}_{i1}$ and $\mathbf{n}_{i2}$ are orthogonal to each other, and perpendicular to $\mathbf{q}_{hi}$. $\mathbf{q}_{hi}$ is the wavevector for each density-wave, and the absolute value is identical for all branches. It also satisfies $\sum_{i=1}^{n} \mathbf{q}_{hi} = 0$. κ_i is the phase for each wave, and usually same for n waves, meaning that the coherent density-waves are cross-propagating. $\mathbf{m}_{\text{net}}$ is the uniform magnetisation, describing the ferromagnetic background that is used to compensate the net magnetisation of the system.

Figure 1.4c shows such structure with a $n = 1$ density-wave. This is essentially the helical lattice structure for a helimagnet in the ground state. Figure 1.4e shows the $n = 4$ density-wave, with a square vortex lattice. Curiously, a magnetic quasi-crystal structure can be constructed by a five-$\mathbf{q}$ structure in this way, as shown in Fig. 1.4i.

For $n = 3$ or $n = 6$, otherwise called a triple-$\mathbf{q}$ structure [34, 64], as shown in Fig. 1.4f–h, a six-fold-symmetric texture is expected. By choosing the appropriate κ, the skyrmion lattice with an individual $N = 1$ skyrmion per unit cell may be present (see Fig. 1.4f and h). However, a random phase may lead to a long-range-ordered magnetisation lattice with completely different texture, as can be seen in Fig. 1.4g. In comparison with the 'standard' skyrmion lattice that is composed of axially-symmetric skyrmions (see Fig. 1.4d), the triple-$\mathbf{q}$ structure has extra six-fold-symmetric fine structures that decorate around the corners of each skyrmion vortex. Moreover, the $\theta(\rho)$ relation for the triple-$\mathbf{q}$ skyrmion lattice is linear, which can be chosen as a radial function approximation for the skyrmion solution in Eq. (1.11).

Starting from the bulk system, we now need to find which spin configuration is favoured at a certain temperature and field, leading to the field-temperature phase diagram [4]. (1) At higher temperature, above T_c, the system is paramagnetic. (2) Below T_c at zero field, a multidomain, helical state exists. The propagation wavevector $\mathbf{q}_h$ is locked by the cubic anisotropy [34, 56]. (3) Increasing the magnetic field above a critical field B_{c1} [58], a finite conical angle ξ is formed. In this state, the conical spin spirals are parallel to the field direction, forming a single-domain state. (4) Increasing the magnetic field further above B_{c2} ξ will eventually be driven to $0°$, reaching the field-polarised state. (5) At finite magnetic field just below T_c, the Gaussian thermal fluctuation adds an extra term to the Hamiltonian of the system, written as [34]:

$$E \approx E_0 + \frac{1}{2} \log \left[\det \left(\frac{\delta^2 E_0}{\delta \mathbf{m} \delta \mathbf{m}} \right) \right], \quad (1.17)$$

where $E_0 = \int w d^3\mathbf{r}$. With this energy modification, a hexagonally-ordered skyrmion lattice, or triple-$\mathbf{q}$ skyrmion crystal solution is more stable than the conical phase. The three-dimensional skyrmion crystal structure is depicted in Fig. 1.3c, in which the two-dimensional skyrmion lattice, that is perpendicular to $\mathbf{B}$, is repeatedly stacked in the third dimension, forming a skyrmion-tube structure. The in-plane skyrmion lattice is locked along a certain direction that is governed by the weak cubic anisotropy, thus making it a single-domain state.

This theoretical framework can accurately describe the magnetism of the bulk $P2_13$ helimagnets, which share extremely similar magnetic phase diagrams [4, 60]. Though being different in the exact value of the critical field and critical temperature, a universal behaviour of the magnetism among all $P2_13$ helimagnets is observed. 'Universal phase diagrams' for several typical materials are shown in Fig. 1.5a, c and e.

If the thickness of the bulk sample is significantly thinned down (thinner than the helical or conical pitch), the conical phase is largely suppressed for field-out-of-plane geometry [6, 64, 78, 79]. In this case, the skyrmion phase is more favoured by the system, leading to largely extended region of the skyrmion phase in the phase diagram, as shown in Fig. 1.5b, d, f and g.

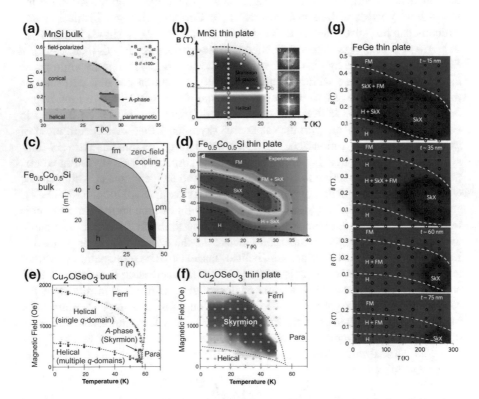

Fig. 1.5 Experimental phase diagrams for typical $P2_13$ helimagnets for both bulk and thin plates samples. **a** Adapted with permission from Ref. [34]. Copyright 2009 The American Association for the Advancement of Science. **b** Adapted with permission from Ref. [76]. Copyright 2012 American Chemical Society. **c** Adapted with permission from Ref. [77]. Copyright 2013 The American Association for the Advancement of Science. **d** Adapted with permission from Ref. [36]. Copyright 2011 Springer Nature. **e–f** Adapted with permission from Ref. [51]. Copyright 2012 The American Association for the Advancement of Science. **g** Adapted with permission from Ref. [39]. Copyright 2011 Springer Nature

1.2 The Properties of Magnetic Skyrmions

Next, we will to discuss some unconventional properties of magnetic skyrmions, arising from their unique topological feature. Though there are a variety of fascinating physical effects concerning the interaction between magnetic skyrmions and other matter, we focus on two interactions that are at the core of current interest. The first one is the conduction-electron-skyrmion interaction, which gives rise to a series of novel magnetoelectric effects. The second one is the magnon-skyrmion interaction, which leads to unique dynamical features.

When dealing with these interactions, a complete Lagrangian (instead of the static $\mathscr{L}$ in Eq. (1.6)) taking into account the time-dependence has to be considered [5]. By adding a dynamical term into Eq. (1.6), the dynamical equation of the magnetisation field is obtained, which essentially reproduces the Landau-Lifshitz equation that is necessary to capture the time-dependence of the system [80]. This also relates to the micromagnetic theory, in which the continuum approximation is applied [80, 81]. The core part of the theory can be written as:

$$\frac{d\mathbf{m}}{dt} = -\gamma \mathbf{m} \times \left(-\frac{\delta E}{\delta \mathbf{m}} \right) , \tag{1.18}$$

where $\gamma = g\mu_B/\hbar$ is the gyromagnetic ratio, and the functional derivative $-\delta E/\delta \mathbf{m}$ stands for the effective field that the local magnetisation 'feels'. In a realistic situation, a damping term is to be added, giving rise to the standard Landau-Lifshitz-Gilbert (LLG) equation [80]. The modification of the dynamical terms can address extra effects. For example, the effective micromagnetic model for the spin transfer torque effect is captured by adding a Slonczewski term (making it the LLGS equation) [82–84]. The thermal fluctuations due to the finite temperature can be addressed by extra stochastic corrections (called SLLG equation) [85, 86].

We now assume that due to some interactions between the skyrmions and other matter, the skyrmion texture undergoes steady-state motion, i.e., its shape is rigid and does not change over time while drifting in real space. Therefore, the solution $\mathbf{m}(\mathbf{r}, t)$ of Eq. (1.18) can be written in the form of $\mathbf{m}(\mathbf{r} - \mathbf{v}_d t)$ [66, 67], where $\mathbf{v}_d$ is the skyrmion drift velocity [87–91]. If we consider the skyrmion vortex as a rigid object, a centre-of-mass coordinate $\mathbf{R} = (R_1, R_2, R_3)$ can be defined, e.g., the vortex core can be the centre-of-mass. Also, $\mathbf{v}_d = d\mathbf{R}/dt$ describes the velocity of this point. The left hand side of Eq. (1.18) becomes [6]:

$$\frac{d\mathbf{m}}{dt} = \frac{d\mathbf{m}(\mathbf{r} - \mathbf{v}_d t)}{dt} = -\sum_{i=1}^{3} \frac{dR_i}{dt} \partial_i \mathbf{m} = -(\mathbf{v}_d \cdot \nabla)\mathbf{m} . \tag{1.19}$$

We then cross multiply $\mathbf{m}$ on both sides of Eq. (1.18), and obtain [6]:

$$-\mathbf{m} \times (\mathbf{v}_d \cdot \nabla)\mathbf{m} = \gamma \left(\mathbf{m} \cdot \frac{\delta E}{\delta \mathbf{m}} \right) \mathbf{m} - \gamma M_s^2 \frac{\delta E}{\delta \mathbf{m}} . \tag{1.20}$$

We then take the inner product of $-\partial_i\mathbf{m}$, and integrate over the two-dimensional real space, obtaining [6]:

$$\int d^2\mathbf{r}\,\partial_i\mathbf{m}\cdot[\mathbf{m}\times(\mathbf{v}_d\cdot\nabla)\mathbf{m}] = \int d^2\mathbf{r}\,\partial_i\mathbf{m}\cdot\left[\gamma\left(\mathbf{m}\cdot\frac{\delta E}{\delta\mathbf{m}}\right)\mathbf{m}\right] - \gamma M_S^2\int d^2\mathbf{r}\,\partial_i\mathbf{m}\cdot\frac{\delta E}{\delta\mathbf{m}}\;,$$

$$-\frac{1}{\gamma M_S^2}v_{dj}\int(\partial_j\mathbf{m}\times\partial_i\mathbf{m})\cdot\mathbf{m}\,d^2\mathbf{r} = 0 - \nabla_R E\;.$$

$$\tag{1.21}$$

Referring to the expression of the winding number in Eq. (1.10), we have

$$\mathbf{G}\times\mathbf{v}_d = \mathbf{F}\;,\tag{1.22}$$

where $\mathbf{G}$ is the gyromagnetic coupling vector. If the skyrmion lies in the x-y-plane, $\mathbf{G} = (0, 0, G_z)$, where $G_z = -\frac{1}{\gamma M_S^2}\int(\partial_j\mathbf{m}\times\partial_i\mathbf{m})\cdot\mathbf{m}\;d^2\mathbf{r} = -4\pi N/(\gamma M_S^2)$, which directly relates to the topological number N. $\mathbf{F} = -\nabla E$ takes the form of a force, and can therefore be regarded as the effective force acting on the skyrmion. Equation (1.22) means that the force acting on a skyrmion equals to the cross product between the topological winding gyromagnetic vector and its velocity. In other words, the skyrmion always drifts perpendicular to the effective force acting on it. This is the Thiele equation of motion [66, 67], and it turns out to be extremely useful for providing intuitive theoretical models describing skyrmion-matter interaction problems, due to its classical mechanics analogy.

1.2.1 Interaction with Conduction Electrons

If the helimagnet is metallic, the coupling between the condition electrons and the localised spins leads to following interaction Hamiltonian [5, 6]:

$$H_i = -J_{ex}\sigma\cdot\mathbf{m}(\mathbf{r}, t)\;,\tag{1.23}$$

where σ is the conduction-electron spin represented by a Pauli matrix, and $\mathbf{m}(\mathbf{r},t)$ is the local magnetisation. For a skyrmion texture, $\mathbf{m}$ takes the form of Eq. (1.11) using polar coordinates. When the coupling of J_{ex} is strong, the conduction electron spin always follows the local magnetisation. Therefore the wave function ϕ of the conduction electron can be written as [92–94]:

$$i\hbar\frac{\partial}{\partial t}\phi = \left(\frac{\mathbf{p}^2}{2m_e} - H_i\right)\phi\;,\tag{1.24}$$

where m_e is the effective electron mass. Let us first define the quantisation axis of the spin angular momentum to be along the z-axis. For a spatially varying magnetisation texture, the conduction electron will adapt the spin to be parallel (or antiparallel) to

$\mathbf{m}(\mathbf{r}, t)$. Consequently, the result of interacting with all $\mathbf{m}(\mathbf{r})$ (simultaneously) can be written as [92]:

$$\Phi = U(\mathbf{r}, t)\phi, \tag{1.25}$$

where

$$U(\mathbf{r}, t) = e^{-i\theta\sigma\cdot\mathbf{n}/2}. \tag{1.26}$$

This is equivalent to applying the rotation operation $-\theta$ about the rotation axis $\mathbf{n} = \frac{\mathbf{e}_z \times \mathbf{m}}{|\mathbf{e}_z \times \mathbf{m}|}$ on the wave function. The Schrödinger equation of [Eq. (1.24)] now reads:

$$i\hbar\frac{\partial}{\partial t}\phi = \left[\frac{(\mathbf{p} + e\mathbf{A})^2}{2m_e} - J_{\text{ex}}\sigma_z - eV\right]\phi, \tag{1.27}$$

where $-e$ is the elementary charge constant. $\mathbf{A}$ and V take the form of the magnetic vector potential operator and electric (scalar) potential operator, respectively, written as [92, 94]:

$$\mathbf{A} = -\frac{i\hbar}{e}U^\dagger\nabla U,$$
$$V = \frac{i\hbar}{e}U^\dagger\partial_t U. \tag{1.28}$$

Assuming that $\mathbf{m}(\mathbf{r}, t)$ varies smoothly and slowly in space and time, all the 'changes' the conduction-electrons experience are adiabatic. Therefore, the $\mathbf{A}$ and V terms can be treated as perturbations:

$$H \approx H_0 + H',$$
$$H_0 = \frac{\mathbf{p}^2}{2m_e} - J_{\text{ex}}\sigma_z. \tag{1.29}$$

The H_0 term is nothing but a spin-up electron (with regards to our quantisation axis) in the nearly-free limit. Evaluating the H' term, it corresponds to an effective electromagnetic field arising from $\mathbf{m}(\mathbf{r}, t)$ that imposes on the conduction electrons. Therefore, the effective magnetic and electric fields take the form of [88, 92–95]:

$$B_i^{\text{em}} = \varepsilon_{ijk}(\partial_j A_k - \partial_k A_j) = \frac{\hbar}{2e}\varepsilon_{ijk}\mathbf{m}\cdot(\partial_j\mathbf{m} \times \partial_k\mathbf{m}),$$
$$E_i^{\text{em}} = -\partial_i V - \partial_t V_i = \frac{\hbar}{e}\mathbf{m}\cdot(\partial_i\mathbf{m} \times \partial_t\mathbf{m}). \tag{1.30}$$

Equation (1.30) also refers to the 'emergent electromagnetic field', as both B_i^{em} and E_i^{em} do not appear to take effect until the conduction electrons start to move through the non-uniform magnetisation pattern, making it an emergent effect. As can be found in Eq. (1.30), the B_i^{em} term contains the expression of the topological number density. Therefore, for an individual skyrmion, the total effective magnetic flux is the spatial integration over a two-dimensional skyrmion vortex area. This leads to [6]

$$\int_{\text{sk}} d^2\mathbf{r}\, B_3^{\text{em}} = -2\pi \frac{\hbar}{e} \ , \tag{1.31}$$

meaning that the emergent flux is quantised by each skyrmion.

We now assume that the electric current with current density $\mathbf{j}$ is applied to a metallic helimagnet, within which a skyrmion lattice exists. This translates into the averaged electron velocity $\mathbf{v}_e$, with the relation of $\mathbf{j} = -e\rho_e\mathbf{v}_e$, where ρ_e is the electron density. Further, the conduction band spin-splitting relates $\mathbf{v}_e$ to the averaged, spin-polarised electron velocity $\mathbf{v}_s$ [6]. The interaction between the conduction electrons and the skyrmion texture will lead to three consequences:

1. The emergent magnetic field that acts on the conduction electrons will gives rise to a fictitious Lorentz force $\mathbf{F} = -e\mathbf{v}_s \times \mathbf{B}^{\text{em}}$. This essentially deflects the electrons, resulting in a Hall effect. Usually this offset is in addition to the ordinary Hall effect (OHE, due to the charge carriers) and anomalous Hall effect (AHE, due to the spin-split conduction band) in a magnetic material [96]. This extra Hall effect is called topological Hall effect (THE) [35, 97, 98]. It therefore leads to the experimental relationship for the measured Hall resistivity ρ_{xy} of [2]

$$\rho_{xy} = \rho_{xy}^{\text{OHE}} + \rho_{xy}^{\text{AHE}} + \rho_{xy}^{\text{THE}} \ , \tag{1.32}$$

   which can be experimentally separated by carrying out field- or temperature dependent Hall measurements, together with magnetometry as a reference for the anomalous Hall effect contribution [35, 40, 42, 44, 88, 97, 99–103].
2. The conduction electrons in turn exchange angular momentum with the local magnetisation through the spin transfer torque effect [82–84]. This will act as a force that impinges on to the skyrmions [87]. Using the LLGS treatment, the Thiele equation in Eq. (1.22) is modified to [5, 6]:

$$\mathbf{F}_{\text{pinning}} = \mathbf{G} \times (\mathbf{v}_s - \mathbf{v}_d) + \mathbf{D}(\beta\mathbf{v}_s - \alpha\mathbf{v}_d) \ , \tag{1.33}$$

   where α and β are damping coefficients, and $\mathbf{D}$ is the dissipative tensor. The force $\mathbf{F}_{\text{pinning}}$ is due to defects or impurities that effectively pin the magnetic texture, preventing the skyrmions from drifting. For a small $\mathbf{v}_s$, the pinning force is relatively strong, giving rise to $\mathbf{v}_d = 0$. For a large $\mathbf{v}_s$ that is above a critical value, a non-zero $\mathbf{v}_d$ can be obtained [90, 104]. In this case, the electric current will drive the collective motion of the skyrmions. The analogous scenario is that of a magnetic domain wall driven by spin transfer torque, arising from the spin-polarised electric current [82]. Nevertheless, it has been demonstrated both theoretically and experimentally that the threshold current that drives the skyrmions to move is several orders of magnitudes smaller than required to drive ferromagnetic domain walls [104–106].
3. While the skyrmions are drifting with velocity $\mathbf{v}_d$, the time-varying quantised magnetic flux will in turn induce an induction, with the electric field written as:

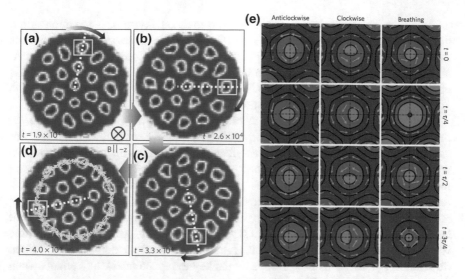

Fig. 1.6 **a–d** SLLG simulation showing the rotation motion of the skyrmion lattice under a radial temperature gradient. It is qualitatively consistent with the LTEM measurements. Reprinted with permission from Ref. [107]. Copyright 2014 by Springer Nature. **e** Three excitation modes can be detected and quantitatively modelled. Figure reproduced from Ref. [108] published under CC-BY license

$$\mathbf{E}^{em} = -\mathbf{v}_d \times \mathbf{B}^{em} \ . \tag{1.34}$$

The effective Lorentz force acting on the conduction electrons due to the emergent fields can therefore be written as [5]:

$$\begin{aligned} \mathbf{F}^{em} &= -e(\mathbf{E}^{em} + \mathbf{v}_s \times \mathbf{B}^{em}) \ , \\ &= -e(\mathbf{v}_s - \mathbf{v}_d) \times \mathbf{B}^{em} \ . \end{aligned} \tag{1.35}$$

This will lead to an extra THE signal, as confirmed by the experiments [88].

1.2.2 Interaction with Magnons

As described by Eq. (1.22), when an incoming flux of magnons impinges on the skyrmions, they transfer momentum to the magnetisation texture. This can be captured by a finite **F**. Interestingly, **F** is not entirely parallel to the propagation direction of the incoming magnons. Instead, the skyrmions tend to be deflected perpendicularly to the magnon propagation direction, analogous to a Magnus force due to the non-trivial topology of the skyrmion structure, i.e., its gyromagnetic coupling vector **G** [5, 91]. This Magnus force also occurs when the skyrmions are pushed by conduction electrons.

An observable experiment that shows the magnon-skyrmion interaction is the thermally-driven skyrmion rotation motion [107]. A radial temperature gradient leads to a magnon current flowing from hot to cold regions [109]. Consequently, the induced **F** forces the skyrmions to rotate around the centre of the temperature gradient. This is observed using LTEM for MnSi and Cu_2OSeO_3 thin plates. The experimental results are consistent with the SLLG simulations, as shown in Fig. 1.6a.

The other unconventional dynamical property of the skyrmion lattice is their ferromagnetic resonance (FMR) behaviour, excited by microwaves [61, 110, 111]. There are three excitation modes that can be detected in FMR experiments, namely clockwise and anticlockwise rotation modes of the individual skyrmions, and breathing modes, as shown in Fig. 1.6e. These results can be quantitatively modelled using a universal dynamical theoretical framework taking the $E_{dipolar}$ into account [61].

While many other appealing properties of magnetic skyrmions are not covered here, the unique features arising from skyrmions interacting with matter may come from the topological argument, i.e., the robust topological winding number written down in Eq. (1.13). This may always be the key to induce interesting effects when dealing with skyrmion-related interactions.

1.3 Characterisation of Magnetic Skyrmions

$P2_13$ helimagnets hosting the skyrmion lattice phase, have individual skyrmion sizes ranging from 3 to 100 nm. This is an inconvenient lengthscale for magnetic characterisation techniques. As a result, not many experimental characterisation methods can be used to reveal the microscopic structure of these skyrmions. In contrast, the larger bubble-skyrmions and artificial skyrmions with dimensions on the $\sim \mu$m scale, can be observed with a wide range of established characterisation techniques [9–12, 19–22, 112].

1.3.1 Magnetic Neutron Diffraction

The first and very suited technique used to characterise magnetic skyrmions was small angle neutron scattering (SANS), performed on MnSi single crystal [34]. The long-range-ordered periodic nature of the skyrmion lattice provides a well-defined reciprocal space lattice of the magnetisation modulation profile. The incommensurate magnetic crystal makes the data interpretation much easier, as no nuclear terms contribute to the magnetic peaks. Neutrons are spin-$\frac{1}{2}$ particles, which carry a magnetic moment of $-\gamma\mu_N$, where $\gamma = 1.913$ is the gyromagnetic ratio. $\mu_N = e\hbar/2m_n$ is the nuclear magneton, and m_n is the neutron mass [113]. The neutron sensitivity to the magnetic structure of the sample comes from the interaction between the neutron spin and the local magnetisation. The magnetic scattering length p that relates to the atomic spin S is given by [113]

$$p = \left(\frac{\mu_0}{4\pi}\right)\left(\frac{e^2}{m_e}\right)\gamma S f(\mathbf{q}) \,, \tag{1.36}$$

where $f(\mathbf{q})$ is the magnetic form factor, $\mathbf{q}$ is the scattering wavevector, and μ_0 the vacuum permeability. Assuming that the incoming neutrons are unpolarised, the magnetic scattering structure factor $\mathbf{F}_m(\mathbf{q})$ from the spin lattice unit cell can be written as:

$$\mathbf{F}_m(\mathbf{q}) = \sum_n f_n(\mathbf{q}) p_n \mathbf{s}_n e^{i\mathbf{q}\cdot\mathbf{r}_n} \,, \tag{1.37}$$

where the summation is performed over the entire unit cell of the magnetic crystal, indexed by the atomic position n. Here the Debye-Waller factor is neglected. $\mathbf{s}_n$ is the magnetic interaction vector, written as:

$$\mathbf{s}_n = \hat{\mathbf{q}}(\hat{\mathbf{q}} \cdot \hat{\mathbf{m}}_n) - \hat{\mathbf{m}}_n \,, \tag{1.38}$$

where $\hat{\mathbf{q}}$ and $\hat{\mathbf{m}}_n$ are the unit vectors along $\mathbf{q}$ and $\mathbf{m}_n$. Therefore, the scattering intensity from $\mathbf{q}$ is $I(\mathbf{q}) = |\mathbf{F}_m(\mathbf{q})|^2$. It is thus clear that $\mathbf{m}_n$ is explicitly contained in the scattering amplitude. Moreover, only the local magnetisation that has a component perpendicular to the scattering wavevector can contribute to the scattering signal. If $\mathbf{m}_n$ is modulated in real space, the magnetic structure will give rise to magnetic diffraction.

The experimental geometry is shown in Fig. 1.7a. The transmission geometry is used in order to study the magnetic peaks around the (0,0,0) reciprocal space origin. The two-dimensional periodic magnetic lattice therefore gives rise to diffraction peaks (by rocking the sample within a few degrees). The helical phase and skyrmion lattice phase can be clearly identified by the diffraction pattern, as shown in Fig. 1.7b and c. The hexagonally ordered skyrmion lattice manifests itself as a six-fold-symmetric pattern, with one pair locked along ⟨110⟩ by weak cubic anisotropy [34].

The transmission SANS geometry is applicable for all $P2_13$ helimagnets, which proved to be a universal technique necessary to confirm the existence of the skyrmion lattice phase [37, 49, 115–119]. Nevertheless, the fact that magnetic neutron scattering is only sensitive to the orthogonal magnetisation component relative to the momentum transfer $\mathbf{q}$ makes it not sensitive enough to probe the complex skyrmion motif.

1.3.2 Lorentz Transmission Electron Microscopy

The first, and accurate interpretation of the six-fold-symmetric SANS pattern of MnSi as being the skyrmion lattice phase was a great achievement. The reason is that a diffraction experiment simply loses the phase information of the real-space magnetisation distribution [113], leaving only the six magnetic peaks for inferring the real-space structure.

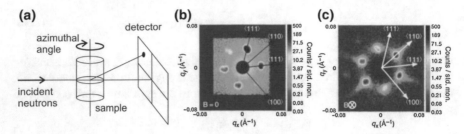

Fig. 1.7 Magnetic neutron diffraction on MnSi. **a** Experimental geometry for SANS. **b** SANS diffraction pattern on helical phase, and **c** skyrmion lattice phase. Panels (**b**) and (**c**) reprinted with permission from Ref. [114]. Copyright 2013 by Elsevier

Naturally, it requires a combination of complementary techniques in order to reveal the real-space information of the skyrmion texture. However, with a modulation periodicity of ∼18 nm in MnSi, there are not too many available imaging techniques available that are able to resolve the structures of this lengthscale. Note that in order to resolve the details inside of the skyrmion vortex, the spatial resolution has to be much smaller than the vortex size. Also, in order to observe the skyrmion lattice, a field of view that contains at least 6–7 skyrmions is required. As it turns out, LTEM is the most suitable imaging technique for these type of measurements.

In an electron microscope, the electrons are accelerated to 100–200 keV, resulting in a wavelength that is much smaller than a typical atomic lattice spacing (e.g., 0.0388 Å for 100 keV electrons). The magnetic sensitivity comes from the Lorentz force when the electron beam interacts with the magnetic specimen [80]. The Lorentz force reads:

$$\mathbf{F}_L = -e(\mathbf{v}_e \times \mathbf{B}) \,, \tag{1.39}$$

where $\mathbf{B}$ is the magnetic flux density. Therefore, the electrons will only be deflected by the $\mathbf{B}$ components that are perpendicular to the electron beam. Note that the Lorentz force has to be integrated over the entire electron path, starting from the electron source, through the specimen, and to the detector. Therefore, it is essential to realise that the stray field (either from the sample or from the beam trajectory) will effectively distort the electron path, giving rise to experimental artefacts. This is usually addressed by applying a systematic sample tilt to compensate for the stray field (i.e., the non-local dipole-dipole interaction).

On the other hand, the electron deflection will not generate image contrast in a bright-field focused image. The non-uniform magnetisation boundary can only be visualised by the shadow effect, i.e., in an out-of-focus scenario. In this defocused mode, the deflection can give contrast on the image plane [80]. As shown in Fig. 1.8, the transmission electron beam can yield a intensity difference if the spins are modulated, and systematically varying from in-plane to out-of-plane [43]. Nevertheless, the high and low intensity regions of the defocused transmission beam only delineates the magnetisation varying boundaries, while the information of the detailed in-plane magnetisation vector distribution is smeared out. This information, on the

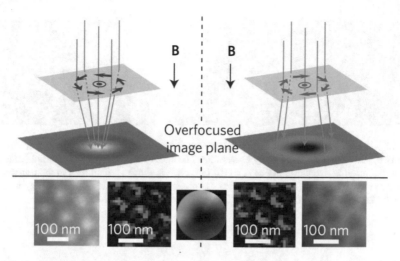

Fig. 1.8 Experimental geometry and representative Lorentz transmission electron images. Detailed discussion of the figure can be found in the main text. Reprinted with permission from Ref. [43]. Copyright 2013 by Springer Nature

other hand, is encoded in the phase of the transmitted electrons [36]. Suppose the beam is emitted along the $\hat{\mathbf{n}}_b$ direction, while $\mathbf{m}_{||} = (m_1, m_2)$ is the in-plane magnetisation vector, then the change of the electron wave's phase φ_e directly relates to the in-plane magnetisation, given by [36]:

$$\nabla_{xy}\varphi_e = -\frac{et}{\hbar}\mathbf{m}_{||} \times \hat{\mathbf{n}}_b . \tag{1.40}$$

Here, t is the thickness of the specimen. This expression specifies that if the phase map of the transmitted electron can be reconstructed, the detailed magnetisation profile (only the in-plane components) can be retrieved. While there are several phase-retrieval methods [120, 121] that are applied in electron microscopy, the most convenient one is to infer the phase distribution by only measuring the intensity distributions. The underlying mechanism is described by the transport-of-intensity equation [36] (Fig. 1.9).

When the propagating electron wave impinges on a specimen, both its amplitude and phase will be distorted. For a phase object, the amplitude change is insignificant and can thus be ignored in this approximation, and also somewhat true for the defocused mode that is used to study the Lorentz deflection [80]. Here, the magnetic contrast does not come from the beam attenuation (amplitude), but from the beam deflection (phase). Under this assumption, one can still expect that the amplitude will increase in some regions and decrease in others, purely due to the phase modulation (wave front deformation) induced by the specimen. The transport-of-intensity links the phase change to the intensity change by:

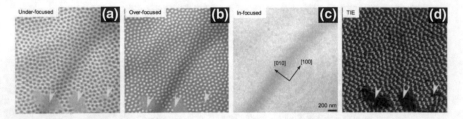

Fig. 1.9 At least two defocused LTEM images on the same site have to be performed, in order to carry out the transport-of-intensity processing. The under-focused (**a**), over-focused (**b**) images for the skyrmion lattice phase in $Fe_{0.5}Co_{0.5}Si$ thin plate is shown, which lead to the transport-of-intensity processed image in (**d**). The in-focused mode (**c**), however, does not show any magnetic contrast. Reprinted with permission from Ref. [36]. Copyright 2010 by Springer Nature

$$\frac{2\pi}{\lambda}\frac{\partial I}{\partial z} = -\nabla_{xy} \cdot (I\nabla_{xy}\varphi_e) , \qquad (1.41)$$

where I is the measured intensity of the transmission electron, and z is the length along the beam path. This is derived from the Schrödinger equation of the high-energy electrons in free space, in which the two real functions I and φ_e replace the complex electron wave function. As a result, the intensity derivative with respect to z has to be performed in order to solve φ_e. This is usually achieved by taking three images with equal spacing, i.e., under-focused $I(z - \delta z)$, in-focus $I(z)$, and over-focused $I(z + \delta z)$ images [122]. Therefore,

$$\frac{I(z + \delta z) - I(z - \delta z)}{2\delta z} = \frac{dI}{dz} + O(\delta z^2) . \qquad (1.42)$$

This simple difference will give a good estimate of the derivative. Consequently, by performing numerical processing using Eqs. (1.41) and (1.40), the lateral magnetisation texture can be solved. Though LTEM is a powerful magnetic microscopy technique that gives microscopic information about skyrmions [36, 39, 43, 45, 47, 51, 76], several aspects are worth noting. First, the transmission mode requires that the specimen is very thin (usually below 200 nm). Therefore, to study bulk magnetism in helimagnets, the sample has to be thinned down to 'nearly two-dimensional' plates. This sample preparation procedure may introduce defects that alter the bulk properties of the material. Second, the transport-of-intensity process is not a direct imaging method that can be unambiguously related to the real-space magnetisation vector distribution. Solving Eq. (1.41) and subsequently solving Eq. (1.40) will involve several approximations and boundary conditions. Eventually, the lateral resolution may be compromised. For example, a typical colour-wheeled skyrmion image obtained from the transport-of-intensity processing only qualitatively shows the in-plane magnetisation components, and it is challenging to retrieve the quantitative information. Further, advanced techniques such as electron holography [121] and differential phase contrast TEM [123] were developed to overcome these limitations, however, the missing m_3 component is a fundamental issue for all TEM-based methods.

1.3.3 Magnetic Force Microscopy

Magnetic force microscopy (MFM) is a variant of atomic force microscopy [80]. Unlike LTEM, MFM is a scanning probe type magnetic imaging technique. In an MFM experiment, magnetic forces are probed via the deflection of the cantilever, which carries the tip-shaped head at its end. Its motion is controlled by piezoelectric actuators, and the deflection of the cantilever due to the tip-sample force is probed by a position-sensitive detector. By carrying out an area scan, the magnetic structure on the surface level can be mapped [34].

In other words, the magnetic tip essentially measures the stray field distribution from the sample, followed by calculations or simulation of the real-space magnetic structure. In an ideal case, assuming the magnetisation distribution $\mathbf{m}_{tip}$ of the tip is known, the measured magnetic force $\mathbf{F}$ relates to the interaction energy E between the tip and the stray field $\mathbf{H}_{sample}$ by [80]:

$$\mathbf{F} = -\nabla E \; ,$$
$$E = -\int_{tip} \mathbf{m}_{tip} \cdot \mathbf{H}_{sample} d^3\mathbf{r} \; . \tag{1.43}$$

Therefore, by analysing the distribution of F on the two-dimensional surface, the stray field distribution from the sample can be obtained, which in turn provides information about the real-space magnetic structure. The magnetic tip is usually magnetised along its axis, i.e., its uniform magnetisation is perpendicular to the sample surface. This allows for the most straightforward data interpretation. As a result, MFM is sensitive to the out-of-plane component of the magnetisation. MFM has been successfully applied to study the skyrmion lattice phase in $Fe_{0.5}Co_{0.5}Si$ [77], Cu_2OSeO_3 [125] (see Fig. 1.10), and GaV_4S_8 [119]. The skyrmion lattice can be seen in the force map, in which the vortices can be easily identified via their m_3 component.

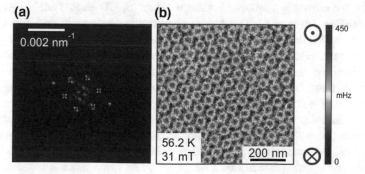

(a) 0.002 nm^{-1} **(b)** 56.2 K 31 mT 200 nm 450 mHz 0

Fig. 1.10 Magnetic force microscopy images of the skyrmion lattice phase of Cu_2OSeO_3. Reprinted with permission from Ref. [124]. Copyright 2016 by American Chemical Society

Compared with LTEM, MFM has several features that can be either an advantage or a disadvantage to image skyrmions. First, due to the scanning probe setup, the probed area in MFM is much more flexible. However, MFM only probes the out-of-plane magnetisation component, which is not sensitive to the m_1 and m_2 components of the skyrmion magnetisation. Measurements of the in-plane magnetisation distribution, however, are always challenging for MFM [80]. Second, it must be taken into account that the magnetism of the sample may be influenced by the tip. This may sometime hamper the data interpretation. However, it also provides an effective way to locally interact with individual skyrmions. Third, the measurement speed is rather slow compared to LTEM, so it does not allow for studying the dynamics of the system on the timescale of seconds in a conventional setup. Last, MFM is a surface-sensitive technique, which limits the probed volume. However, it provides more precise information about a near-surface region [80].

It is worth noting that another scanning probe microscopy technique, spin-polarised scanning tunnelling microscopy (STM) has been successfully used to characterise the type I-(2) surface-induced skyrmion structures [3, 7, 126]. The resolution of STM reaches atomic level, and it is able to resolve all three components of the classical spins. Therefore, STM is ideally suited for studying the atomic-sized skyrmions in two-dimensional materials, such as the epitaxially-grown magnetic thin films with a few monolayers [7]. Nevertheless, the demanding sample environment (e.g., ultra-high vacuum and extremely well-prepared surface), as well as the small probing area (typically less than 1 μm^2) make STM less suited to study the magnetic structures in $P2_13$ helimagnets.

1.4   What Is Missing?

So far, we have summarised the core aspects of magnetic skyrmions in noncentrosymmetric helimagnets. From an experimental point of view, the direct characterisation of magnetic skyrmions is the most important step for studying this fascinating topic. However, the currently available techniques (SANS, LTEM and MFM) have their own limitations, while some of the key bits of information about the detailed skyrmion structures are still missing or at least ambiguous.

Lengthscale-wise, the real-space information with regards to the skyrmion phase can be classified on three levels. (1) On the macroscopic scale, the long-range-ordered skyrmion lattice phase appears as a single domain state. The correlation length reaches the mm range [127]. This can be probed by neutrons. However, LTEM and MFM cannot achieve such large fields of view. (2) On the mesoscopic scale, the skyrmion lattice may break up into domains, with the typical domain size ranging from several to hundreds of micrometers [124, 128]. The distribution, size, and shape of the skyrmion lattice domains, are of great importance for understanding the system. However, neither of these three methods can effectively provide this piece of information. (3) On the microscopic scale, the individual skyrmion carries a fine internal structure, described by Eq. (1.11). As a result, the topological winding number N and the helicity angle χ are the key parameters that govern the topological and chiral properties of the skyrmions. Nevertheless, these two quantities are not directly measurable with the existing techniques.

Dimensionality-wise, two aspects are of great importance in revealing the complete structural information of magnetic skyrmions. (1) The skyrmion crystal is always a three-dimensional object, though the main lateral features can be described by a two-dimensional lattice. In certain circumstances, the three-dimensional skyrmion structure may behave very differently [45, 77, 117, 129]. Therefore, the experimental technique that can provide a depth sensitivity is of great importance from this perspective. This is only possible in dedicated LTEM experiments [121], while being challenging for SANS and MFM. (2) The surface magnetic structure may be very different from the one in the bulk [124, 130–134]. Therefore, a surface-sensitive technique is required to study skyrmion structures on the surface level. From this point of view, MFM is a suitable technique. However, SANS and LTEM are not able to distinguish the surface signal from the overall transmission signals coming from the bulk of the sample.

Furthermore, three different timescales are required to study the interesting phenomena arising in skyrmion systems. (1) On the longest timescale, the static skyrmion phase is to be characterised in detail, especially, the decay phenomena of the metastable state [77]. (2) For slow dynamics, such as the skyrmion phase response to an AC field [58, 129, 135], as well as the thermally-driven rotation motion [107], the characterisation technique is required to have a time resolution of the order of ms. (3) For fast dynamics, such as microwave-driven excitations [61], the time resolution is required to be on the order of less than 1 ns.

These challenges in studying the missing aspects of skyrmions form the motivation of my work. As will be shown in Chaps. 2–5, by only using soft x-rays, the structural information of skyrmions, scattered within the length scales that spans five orders of magnitude (from tens of nm to mm) can be unambiguously characterised. Moreover, it covers the dimensionality issue (depth sensitivity) and the timescale issue (ultrafast detection). These advantages broaden the horizon of skyrmion science and allow us to study magnetic skyrmion systems in more detail, promising many opportunities for the future of this field.

In this thesis, the focus is on the experimental technique of resonant soft x-ray scattering. As only the elastic scattering process is considered and measured, it is called resonant elastic x-ray scattering (REXS) [136].

1.5 Structure of the Thesis

In this chapter, the well-established theory for skyrmions is introduced, reconstructing the picture from single skyrmions to the skyrmion crystal. A few comments about the current characterisation techniques are given. In Chap. 2, we will start with the largest lengthscale, the long-range-ordered skyrmion lattice phase. This is an intensely studied phase, mostly using neutron diffraction, and we will show that this piece of information can be equivalently (or actually even better) obtained using resonant x-ray diffraction. The theoretical foundation of this technique is also given. In Chap. 3, we will demonstrate imaging technique with which we were able to effectively map the skyrmion domains. The measurements also suggest a way to control the formation of skyrmion domains, which might be the key for enabling skyrmion-based device applications. Chapters 4 and 5 present the highlights of this work, in which we will show that using the dichroism extinction rule, the topological winding number and the skyrmion helicity angle can be unambiguously determined. In this sense, this technique is capable of accurately measuring the internal structure of single skyrmions. The outline of this work is summarised in the following figure.

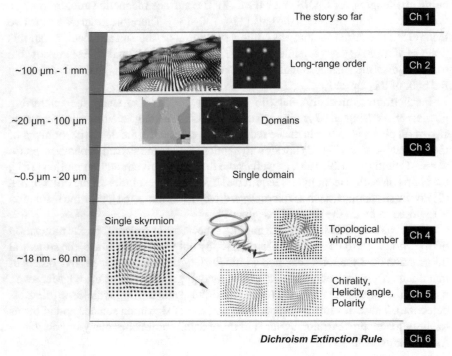

References

1. N.D. Mermin, Rev. Mod. Phys. **51**, 591 (1979)
2. N. Nagaosa, Y. Tokura, Nat. Nanotech. **8**, 899 (2013)
3. S. Heinze, K. von Bergmann, M. Menzel, J. Brede, A. Kubetzka, R. Wiesendanger, G. Bihlmayer, S. Blügel, Nat. Phys. **7**, 713 (2011)
4. A. Bauer, C. Pfleiderer, Topological structures in ferroic materials: domain walls, vortices and skyrmions: generic aspects of skyrmion lattices in chiral magnets, in *Topological Structures in Ferroic Materials: Domain Walls, Vortices and Skyrmions* (Springer International Publishing, Cham, 2016), pp. 1–28

5. M. Garst, Topological structures in ferroic materials: domain walls, vortices and skyrmions: topological skyrmion dynamics in chiral magnets, in *Topological Structures in Ferroic Materials: Domain Walls, Vortices and Skyrmions* (Springer International Publishing, Cham, 2016), pp. 29–53

6. S. Seki, M. Mochizuki, *Skyrmions in Magnetic Materials* (Springer, 2016)

7. R. Wiesendanger, Nat. Rev. Mater. **1**, 16044 (2016)

8. J. Sampaio, V. Cros, S. Rohart, A. Thiaville, A. Fert, Nat. Nanotech. **8**, 839 (2013)

9. W. Jiang, P. Upadhyaya, W. Zhang, G. Yu, M.B. Jungfleisch, F.Y. Fradin, J.E. Pearson, Y. Tserkovnyak, K.L. Wang, O. Heinonen, S.G.E. te Velthuis, A. Hoffmann, Science **349**, 283 (2015)

10. F. Büttner, C. Moutafis, M. Schneider, B. Krüger, C.M. Gunther, J. Geilhufe, C.V. Korff Schmising, J. Mohanty, B. Pfau, S. Schaffert, A. Bisig, M. Foerster, T. Schulz, C.A.F. Vaz, J.H. Franken, H.J.M. Swagten, M. Kläui, S. Eisebitt, Nat. Phys. **11**, 225 (2015)

11. O. Boulle, J. Vogel, H. Yang, S. Pizzini, D. de Souza Chaves, A. Locatelli, T.O. Mentes, A. Sala, L.D. Buda-Prejbeanu, O. Klein, M. Belmeguenai, Y. Roussigné, A. Stashkevich, S.M. Chérif, L. Aballe, M. Foerster, M. Chshiev, S. Auffret, I.M. Miron, G. Gaudin, Nat. Nanotech. **11**, 449 (2016)

12. S. Woo, K. Litzius, B. Krüger, M.-Y. Im, L. Caretta, K. Richter, M. Mann, A. Krone, R.M. Reeve, M. Weigand, P. Agrawal, I. Lemesh, M.-A. Mawass, P. Fischer, M. Kläui, G.S.D. Beach, Nat. Mater. **15**, 501 (2016)

13. C. Moreau-Luchaire, C. Moutafis, N. Reyren, J. Sampaio, C.A.F. Vaz, N.V. Horne, K. Bouze-houane, K. Garcia, C. Deranlot, P. Warnicke, P. Wohlhüter, J.M. George, M. Weigand, J. Raabe, V. Cros, A. Fert, Nat. Nanotechnol. **11**, 444 (2016)

14. X.Z. Yu, M. Mostovoy, Y. Tokunaga, W. Zhang, K. Kimoto, Y. Matsui, Y. Kaneko, N. Nagaosa, Y. Tokura, Proc. Natl. Acad. Sci. U.S.A. **109**, 8856 (2012)

15. X.Z. Yu, Y. Tokunaga, Y. Kaneko, W.Z. Zhang, K. Kimoto, Y. Matsui, Y. Taguchi, Y. Tokura, Nat. Commun. **5**, 3198 (2014)

16. X.Z. Yu, K. Shibata, W. Koshibae, Y. Tokunaga, Y. Kaneko, T. Nagai, K. Kimoto, Y. Taguchi, N. Nagaosa, Y. Tokura, Phys. Rev. B **93**, 134417 (2016)

17. W. Wang, Y. Zhang, G. Xu, L. Peng, B. Ding, Y. Wang, Z. Hou, X. Zhang, X. Li, E. Liu, S. Wang, J. Cai, F. Wang, J. Li, F. Hu, G. Wu, B. Shen, X.-X. Zhang, Adv. Mater. **28**, 6887 (2016)

18. E.A. Giess, Science **208**, 938 (1980)

19. L. Sun, R.X. Cao, B.F. Miao, Z. Feng, B. You, D. Wu, W. Zhang, A. Hu, H.F. Ding, Phys. Rev. Lett. **110**, 167201 (2013)

20. B.F. Miao, L. Sun, Y.W. Wu, X.D. Tao, X. Xiong, Y. Wen, R.X. Cao, P. Wang, D. Wu, Q.F. Zhan, B. You, J. Du, R.W. Li, H.F. Ding, Phys. Rev. B **90**, 174411 (2014)

21. J. Li, A. Tan, K. Moon, A. Doran, M.A. Marcus, A.T. Young, E. Arenholz, S. Ma, R.F. Yang, C. Hwang, Z.Q. Qiu, Nat. Commun. **5**, 4704 (2014)

22. D.A. Gilbert, B.B. Maranville, A.L. Balk, B.J. Kirby, P. Fischer, D.T. Pierce, J. Unguris, J.A. Borchers, K. Liu, Nat. Commun. **6**, 8462 (2015)

23. T.H.R. Skyrme, Proc. Roy. Soc. Lond. A **260**, 127 (1961)

24. T.H.R. Skyrme, Nucl. Phys. **31**, 556 (1962)

25. R.A. Battye, N.S. Manton, P.M. Sutcliffe, Proc. Roy. Soc. Lond. A **463**, 261 (2007)

26. N. Manton, *Topological Solitons*. Lecture notes for the XIII Saalburg summer school (2007)

27. G.E. Brown, M. Rho (eds.), *The Multifaceted Skyrmion* (World Scientific Publishing, 2010)

28. C. Naya, P. Sutcliffe, J. High Energy Phys. **2018**, 174 (2018)

29. C. Pfleiderer, S.R. Julian, G.G. Lonzarich, Nature **414**, 427 (2001)

30. C. Pfleiderer, D. Reznik, L. Pintschovius, H.V. Löhneysen, M. Garst, A. Rosch, Nature **427**, 227 (2004)

31. M. Uchida, Y. Onose, Y. Matsui, Y. Tokura, Science **311**, 359 (2006)

32. C. Pfleiderer, P. Böni, T. Keller, U.K. Rößler, A. Rosch, Science **316**, 1871 (2007)

33. M. Uchida, N. Nagaosa, J.P. He, Y. Kaneko, S. Iguchi, Y. Matsui, Y. Tokura, Phys. Rev. B **77**, 184402 (2008)

34. S. Mühlbauer, B. Binz, F. Jonietz, C. Pfleiderer, A. Rosch, A. Neubauer, R. Georgii, P. Böni, Science **323**, 915 (2009)
35. A. Neubauer, C. Pfleiderer, B. Binz, A. Rosch, R. Ritz, P.G. Niklowitz, P. Böni, Phys. Rev. Lett. **102**, 186602 (2009)
36. X.Z. Yu, Y. Onose, N. Kanazawa, J.H. Park, J.H. Han, Y. Matsui, N. Nagaosa, Y. Tokura, Nature **465**, 901 (2010)
37. W. Münzer, A. Neubauer, T. Adams, S. Mühlbauer, C. Franz, F. Jonietz, R. Georgii, P. Böni, B. Pedersen, M. Schmidt, A. Rosch, C. Pfleiderer, Phys. Rev. B **81**, 041203(R) (2010)
38. A. Bauer, A. Neubauer, C. Franz, W. Münzer, M. Garst, C. Pfleiderer, Phys. Rev. B **82** (2010)
39. X.Z. Yu, N. Kanazawa, Y. Onose, K. Kimoto, W.Z. Zhang, S. Ishiwata, Y. Matsui, Y. Tokura, Nat. Mater. **10**, 106 (2011)
40. N. Kanazawa, Y. Onose, T. Arima, D. Okuyama, K. Ohoyama, S. Wakimoto, K. Kakurai, S. Ishiwata, Y. Tokura, Phys. Rev. Lett. **106**, 156603 (2011)
41. R. Ritz, M. Halder, M. Wagner, C. Franz, A. Bauer, C. Pfleiderer, Nature **497**, 231 (2013)
42. R. Ritz, M. Halder, C. Franz, A. Bauer, M. Wagner, R. Bamler, A. Rosch, C. Pfleiderer, Phys. Rev. B **87**, 134424 (2013)
43. K. Shibata, X.Z. Yu, T. Hara, D. Morikawa, N. Kanazawa, K. Kimoto, S. Ishiwata, Y. Matsui, Y. Tokura, Nat. Nanotech. **8**, 723 (2013)
44. C. Franz, F. Freimuth, A. Bauer, R. Ritz, C. Schnarr, C. Duvinage, T. Adams, S. Blügel, A. Rosch, Y. Mokrousov, C. Pfleiderer, Phys. Rev. Lett. **112**, 186601 (2014)
45. T. Tanigaki, K. Shibata, N. Kanazawa, X. Yu, Y. Onose, H.S. Park, D. Shindo, Y. Tokura, Nano Lett. **15**, 5438 (2015)
46. M. Nagao, Y.-G. So, H. Yoshida, T. Nagai, K. Edagawa, K. Saito, T. Hara, A. Yamazaki, K. Kimoto, Appl. Phys. Express **8**, 033001 (2015)
47. T. Matsumoto, Y.-G. So, Y. Kohno, H. Sawada, Y. Ikuhara, N. Shibata, Sci. Adv. **2**, e1501280 (2016)
48. J.-W.G. Bos, C.V. Colin, T.T.M. Palstra, Phys. Rev. B **78**, 094416 (2008)
49. T. Adams, A. Chacon, M. Wagner, A. Bauer, G. Brandl, B. Pedersen, H. Berger, P. Lemmens, C. Pfleiderer, Phys. Rev. Lett. **108**, 237204 (2012)
50. J.H. Yang, Z.L. Li, X.Z. Lu, M.-H. Whangbo, S.-H. Wei, X.G. Gong, H.J. Xiang, Phys. Rev. Lett. **109**, 107203 (2012)
51. S. Seki, X.Z. Yu, S. Ishiwata, Y. Tokura, Science **336**, 198 (2012)
52. O. Janson, I. Rousochatzakis, A.A. Tsirlin, M. Belesi, A.A. Leonov, U.K. Rößler, J. van den Brink, H. Rosner, Nat. Commun. **5**, 5376 (2014)
53. V.A. Chizhikov, V.E. Dmitrienko, J. Magn. Magn. Mater. **382**, 142 (2015)
54. I. Dzyaloshinskii, J. Phys. Chem. Solids **4**, 241 (1958)
55. T. Moriya, Phys. Rev. **120**, 91 (1960)
56. P. Bak, M.H. Jensen, J. Phys. C: Solid State Phys. **13**, L881 (1980)
57. M. Janoschek, M. Garst, A. Bauer, P. Krautscheid, R. Georgii, P. Böni, C. Pfleiderer, Phys. Rev. B **87**, 134407 (2013)
58. A. Bauer, C. Pfleiderer, Phys. Rev. B **85**, 214418 (2012)
59. A. Bauer, M. Garst, C. Pfleiderer, Phys. Rev. Lett. **110**, 177207 (2013)
60. S. Buhrandt, L. Fritz, Phys. Rev. B **88**, 195137 (2013)
61. T. Schwarze, J. Waizner, M. Garst, A. Bauer, I. Stasinopoulos, H. Berger, C. Pfleiderer, D. Grundler, Nat. Mater. **14**, 478 (2015)
62. A.N. Bogdanov, D.A. Yablonskii, Sov. Phys. JETP **68**, 1 (1989)
63. A.N. Bogdanov, A. Hubert, J. Magn. Magn. Mater. **138**, 255 (1994)
64. S.D. Yi, S. Onoda, N. Nagaosa, J.H. Han, Phys. Rev. B **80**, 054416 (2009)
65. H.-B. Braun, Adv. Phys. **61**, 1 (2012)
66. A.A. Thiele, Phys. Rev. Lett. **30**, 230 (1973)
67. A.A. Thiele, J. Appl. Phys. **45**, 377 (1974)
68. N. Papanicolaou, P.N. Spathis, Nonlinearity **12**, 285 (1999)
69. S. Komineas, N. Papanicolaou, Phys. D **99**, 81 (1996)
70. S.L. Zhang, A.A. Baker, S. Komineas, T. Hesjedal, Sci. Rep. **5**, 15773 (2015)

71. S. Komineas, N. Papanicolaou, Nonlinearity **11**, 265 (1998)
72. U.K. Rößler, A.A. Leonov, A.N. Bogdanov, J. Phys. Conf. Ser. **303**, 012105 (2011)
73. U.K. Rößler, A.N. Bogdanov, C. Pfleiderer, Nature **442**, 797 (2006)
74. Y. Tokura, S. Seki, Adv. Mater. **22**, 1554 (2010)
75. Y. Tokura, S. Seki, N. Nagaosa, Rep. Prog. Phys. **77**, 076501 (2014)
76. A. Tonomura, X.Z. Yu, K. Yanagisawa, T. Matsuda, Y. Onose, N. Kanazawa, H.S. Park, Y. Tokura, Nano Lett. **12**, 1673 (2012)
77. P. Milde, D. Köhler, J. Seidel, L.M. Eng, A. Bauer, A. Chacon, J. Kindervater, S. Mühlbauer, C. Pfleiderer, S. Buhrandt, C. Schütte, A. Rosch, Science **340**, 1076 (2013)
78. F.N. Rybakov, A.B. Borisov, A.N. Bogdanov, Phys. Rev. B **87**, 094424 (2013)
79. A.O. Leonov, Y. Togawa, T.L. Monchesky, A.N. Bogdanov, J. Kishine, Y. Kousaka, M. Miyagawa, T. Koyama, J. Akimitsu, T. Koyama, K. Harada, S. Mori, D. McGrouther, R. Lamb, M. Krajnak, S. McVitie, R.L. Stamps, K. Inoue, Phys. Rev. Lett. **117**, 8 (2016)
80. A. Hubert, R. Schäfer, *Magnetic Domains—The Analysis of Magnetic Microstructures* (Springer, 2008)
81. S.J. Blundell, *Magnetism in Condensed Matter* (Oxford University Press, 2001)
82. J.C. Slonczewski, J. Magn. Magn. Mater. **159**, L1 (1996)
83. L. Berger, Phys. Rev. B **54**, 9353 (1996)
84. S. Zhang, Z. Li, Phys. Rev. Lett. **93**, 127204 (2004)
85. W.F. Brown, Phys. Rev. **130**, 1677 (1963)
86. K. Miyazakia, K. Seki, J. Chem. Phys. **108**, 7052 (1998)
87. K. Everschor, M. Garst, B. Binz, F. Jonietz, S. Mühlbauer, C. Pfleiderer, A. Rosch, Phys. Rev. B **86**, 054432 (2012)
88. T. Schulz, R. Ritz, A. Bauer, M. Halder, M. Wagner, C. Franz, C. Pfleiderer, K. Everschor, M. Garst, A. Rosch, Nat. Phys. **8**, 301 (2012)
89. J. Iwasaki, M. Mochizuki, N. Nagaosa, Nat. Nanotech. **8**, 742 (2013)
90. J. Iwasaki, M. Mochizuki, N. Nagaosa, Nat. Commun. **4**, 1463 (2013)
91. C. Schütte, M. Garst, Phys. Rev. B **90**, 094423 (2014)
92. G.E. Volovik, J. Phys. C. Solid State Phys. **20**, L83 (1987)
93. N. Nagaosa, X.Z. Yu, Y. Tokura, Phil. Trans. Roy. Soc. A **370**, 5806 (2012)
94. N. Nagaosa, Y. Tokura, Phys. Scr. **T146**, 014020 (2012)
95. J. Zang, M. Mostovoy, J.H. Han, N. Nagaosa, Phys. Rev. Lett. **107**, 136804 (2011)
96. N. Nagaosa, J. Sinova, S. Onoda, A.H. MacDonald, N.P. Ong, Rev. Mod. Phys. **82**, 1539 (2010)
97. Y. Taguchi, Y. Oohara, H. Yoshizawa, N. Nagaosa, Y. Tokura, Science **291**, 2573 (2001)
98. F.D.M. Haldane, Phys. Rev. Lett. **93**, 206602 (2004)
99. M. Lee, W. Kang, Y. Onose, Y. Tokura, N.P. Ong, Phys. Rev. Lett. **102**, 186601 (2009)
100. S.X. Huang, C.L. Chien, Phys. Rev. Lett. **108**, 267201 (2012)
101. Y. Li, N. Kanazawa, X.Z. Yu, A. Tsukazaki, M. Kawasaki, M. Ichikawa, X.F. Jin, F. Kagawa, Y. Tokura, Phys. Rev. Lett. **110**, 117202 (2013)
102. T. Yokouchi, N. Kanazawa, A. Tsukazaki, Y. Kozuka, M. Kawasaki, M. Ichikawa, F. Kagawa, Y. Tokura, Phys. Rev. B **89**, 064416 (2014)
103. N. Kanazawa, M. Kubota, A. Tsukazaki, Y. Kozuka, K.S. Takahashi, M. Kawasaki, M. Ichikawa, F. Kagawa, Y. Tokura, Phys. Rev. B **91**, 041122(R) (2015)
104. K. Everschor, M. Garst, R.A. Duine, A. Rosch, Phys. Rev. B **84**, 064401 (2011)
105. F. Jonietz, S. Mühlbauer, C. Pfleiderer, A. Neubauer, W. Münzer, A. Bauer, T. Adams, R. Georgii, P. Böni, R.A. Duine, K. Everschor, M. Garst, A. Rosch, Science **330**, 1648 (2010)
106. X.Z. Yu, N. Kanazawa, W.Z. Zhang, T. Nagai, T. Hara, K. Kimoto, Y. Matsui, Y. Onose, Y. Tokura, Nat. Commun. **3**, 988 (2012)
107. M. Mochizuki, X.Z. Yu, S. Seki, N. Kanazawa, W. Koshibae, J. Zang, M. Mostovoy, Y. Tokura, N. Nagaosa, Nat. Mater. **13**, 241 (2014)
108. M. Garst, J. Waizner, D. Grundler, J. Phys. D Appl. Phys. **50**, 293002 (2017)
109. L. Kong, J. Zang, Phys. Rev. Lett. **111**, 067203 (2013)
110. M. Mochizuki, Phys. Rev. Lett. **108**, 017601 (2012)

111. Y. Onose, Y. Okamura, S. Seki, S. Ishiwata, Y. Tokura, Phys. Rev. Lett. **109**, 037603 (2012)
112. G. Chen, A. Mascaraque, A.T. N'Diaye, A.K. Schmid, Appl. Phys. Lett. **106**, 242404 (2015)
113. B.T.M. Willis, C.J. Carlile, *Experimental Neutron Scattering* (Oxford University Press, 2013)
114. D.L. Price, F. Fernandez-Alonso, in *Neutron Scattering–Fundamentals*, vol. 44, ed. by F. Fernandez-Alonso, D.L. Price. Experimental Methods in the Physical Sciences (Academic Press, 2013), pp. 1–136
115. T. Adams, S. Mühlbauer, A. Neubauer, W. Münzer, F. Jonietz, R. Georgii, B. Pedersen, P. Böni, A. Rosch, C. Pfleiderer, J. Phys. Conf. Ser. **200**, 032001 (2010)
116. S. Seki, J.-H. Kim, D.S. Inosov, R. Georgii, B. Keimer, S. Ishiwata, Y. Tokura, Phys. Rev. B **85**, 220406(R) (2012)
117. N. Kanazawa, J.-H. Kim, D.S. Inosov, J.S. White, N. Egetenmeyer, J.L. Gavilano, S. Ishiwata, Y. Onose, T. Arima, B. Keimer, Y. Tokura, Phys. Rev. B **86**, 134425 (2012)
118. Y. Tokunaga, X.Z. Yu, J.S. White, H.M. Ronnow, D. Morikawa, Y. Taguchi, Y. Tokura, Nat. Commun. **6**, 7638 (2015)
119. I. Kezsmarki, S. Bordacs, P. Milde, E. Neuber, L.M. Eng, J.S. White, H.M. Ronnow, C.D. Dewhurst, M. Mochizuki, K. Yanai, H. Nakamura, D. Ehlers, V. Tsurkan, A. Loidl, Nat. Mater. **14**, 1116 (2015)
120. M. Khan (ed.) *The Transmission Electron Microscope* (InTech, 2012)
121. H.S. Park, X.Z. Yu, S. Aizawa, T. Tanigaki, T. Akashi, Y. Takahashi, T. Matsuda, N. Kanazawa, Y. Onose, D. Shindo, A. Tonomura, Y. Tokura, Nat. Nanotech. **9**, 337 (2014)
122. K. Ishizuka, B. Allman, J. Elec. Micr. **54**, 171 (2005)
123. D. McGrouther, R.J. Lamb, M. Krajnak, S. McFadzean, S. McVitie, R.L. Stamps, A.O. Leonov, A.N. Bogdanov, Y. Togawa, New J. Phys. **18**, 095004 (2016)
124. S.L. Zhang, A. Bauer, D.M. Burn, P. Milde, E. Neuber, L.M. Eng, H. Berger, C. Pfleiderer, G. van der Laan, T. Hesjedal, Nano Lett. **16**, 3285 (2016)
125. P. Milde, E. Neuber, A. Bauer, C. Pfleiderer, H. Berger, L.M. Eng, Nano Lett. **16**, 5612 (2016)
126. N. Romming, C. Hanneken, M. Menzel, J.E. Bickel, B. Wolter, K. von Bergmann, A. Kubetzka, R. Wiesendanger, Science **341**, 636 (2013)
127. T. Adams, S. Mühlbauer, C. Pfleiderer, F. Jonietz, A. Bauer, A. Neubauer, R. Georgii, P. Böni, U. Keiderling, K. Everschor, M. Garst, A. Rosch, Phys. Rev. Lett. **107**, 217206 (2011)
128. J. Rajeswaria, H. Pinga, G.F. Mancini, Y. Murooka, T. Latychevskaia, D. McGrouther, M. Cantoni, E. Baldini, J.S. White, A. Magrez, T. Giamarchi, H.M. Rønnow, F. Carbone, Proc. Natl. Acad. Sci. U.S.A. **112**, 14212 (2015)
129. J.K.S. Mühlbauer, T. Adams, A. Bauer, U. Keiderling, C. Pfleiderer, New J. Phys. **18**, 075017 (2016)
130. M.N. Wilson, E.A. Karhu, A.S. Quigley, U.K. Roßler, A.B. Butenko, A.N. Bogdanov, M.D. Robertson, T.L. Monchesky, Phys. Rev. B **86**, 144420 (2012)
131. E.A. Karhu, U.K. Rößlerler, A.N. Bogdanov, S. Kahwaji, B.J. Kirby, H. Fritzsche, M.D. Robertson, C.F. Majkrzak, T.L. Monchesky, Phys. Rev. B **85**, 094429 (2012)
132. M.N. Wilson, E.A. Karhu, D.P. Lake, A.S. Quigley, A.N. Bogdanov, U.K. Rößler, T.L. Monchesky, Phys. Rev. B **88**, 214420 (2013)
133. S.A. Meynell, M.N. Wilson, H. Fritzsche, A.N. Bogdanov, T.L. Monchesky, Phys. Rev. B **90**, 014406 (2014)
134. M.N. Wilson, A.B. Butenko, A.N. Bogdanov, T.L. Monchesky, Phys. Rev. B **89**, 094411 (2014)
135. J.S. White, K. Prša, P. Huang, A.A. Omrani, I. Živkovi ć, M. Bartkowiak, H. Berger, A. Magrez, J. L. Gavilano, G. Nagy, J. Zang, H.M. Rønnow. Phys. Rev. Lett. **113**, 107203 (2014)
136. S.W. Lovesey, S.P. Collins, *X-ray Scattering and Absorption by Magnetic Materials* (Oxford University Press, 2002)

Chapter 2
Measurement of the Magnetic Long-Range Order

The structural determination by x-ray diffraction is a well-established technique, utilising the interaction between electromagnetic waves and the charge density of the electrons, called Thomson scattering [1]. The scattering process is clearly taking place between the photons and electrons, implying the sensitivity to the other two degrees of freedom of electrons, i.e., the orbitals and spins are accessible as well. The history of the studies on these interactions, reviewed in [2], can be traced back to the 1970s, when Platzman and Tzoar first showed the sensitivity of the x-ray scattering amplitude to magnetism in theory [3], followed by the experimental proof by DeBergevin and Brunel [4]. This technique is termed as non-resonant magnetic scattering [5]. Nevertheless, the magnetic scattering cross-section is so weak that it was almost invisible in experiments at that time [6].

This situation changed drastically in 1980s, when synchrotron radiation sources allowed the photon energy to be tuned in a controlled way [1]. The celebrated work by Thole and van der Laan showed the strong polarisation dependence of the x-ray electric dipole interaction from the absorption spectrum, when the transition occurs from a core level to the magnetically polarised valence shell [7, 8]. This is the fundamental mechanism by which photons 'feel' the magnetic moment during absorption. Consequently, when the photons are re-emitted, the similar, yet reverse process occurs, in which the same magnetic information is encoded. This two-step process is virtually equivalent to the elastic scattering process (at resonance), and it specifies the physical mechanism for REXS to probe the magnetic moment density.

The milestone was set by Blume, Gibbs et al. [9, 10] and Hannon et al. [11], who did pioneering work on resonant x-ray scattering, and found the exact polarisation dependence of the process. After almost three decades, this technique is mature and suitable to study many magnetic materials, thanks to the fast development of synchrotron radiation. Though the nomenclature may differ in the literature, in this work we call this process resonant elastic x-ray scattering, or REXS. For $3d$ transition metals at the $L_{2,3}$ edges, and rare earth metals at the $M_{4,5}$ edges, the resonant signal due to magnetism is comparable with the Thomson scattering, therefore providing

© Springer Nature Switzerland AG 2018
S. Zhang, *Chiral and Topological Nature of Magnetic Skyrmions*,
Springer Theses, https://doi.org/10.1007/978-3-319-98252-6_2

excellent conditions for actually measuring this process [12, 13]. These resonance conditions require a photon energy range of 0.4–2 keV, which is the soft x-rays spectrum. For this work, we will only concentrate on $3d$ transition metals and their compounds, making use of the L-edge dipole transition. As will be shown, for a dipole transition, the complexity of the scattering process can be largely reduced, making the final analytical form simple and elegant.

2.1 Writing Down the Scattering Process

The periodic magnetic structures of $P2_13$ materials, namely Cu_2OSeO_3, MnSi and FeGe, will be examined by REXS. Therefore, to illustrate the theory, we will use the parameters of Cu_2OSeO_3 as an example, and build up the REXS process from the very elemental mechanisms. One can easily extend this process to other $3d$-compounds and other skyrmion-carrying materials.

There are two points worth considering before applying this technique. First, as the photon energy falls into the range of 400–2000 eV, leading to relatively long wavelengths, the number of accessible materials for experiments in reflection geometry are limited. At the Cu L_3 edge, the x-ray wavelength is 13.3 Å, which is almost the shortest among the $3d$ metals (note that the d^{10} Zn usually does not carry a magnetic moment). Even so, for B20 helimagnetic metals such as MnSi, $Fe_{1-x}Co_xSi$, or FeGe, the (structural) lattice constant is around 4.5–4.7 Å [14], which means that no structural Bragg reflection is accessible. Cu_2OSeO_3, on the other hand, has a relatively large lattice constant (8.925 Å) [15]. This makes it the only accessible material to carry out reflection REXS. For the remaining B20 materials, extra tricks have to be performed.

Second, the penetration depth of soft x-rays varies significantly for energies across the absorption edge [2]. Scattering is more bulk-sensitive for photon energies below the L_3 and further above the L_2 edges, whereas it is more surface-sensitive at resonance. In Cu_2OSeO_3, the x-ray attenuation length for normal incidence at the L_3 edge maximum is 95 nm, while below the absorption edge the attenuation length is 394 nm [16]. Therefore, the magnetic structure we probe is near the surface. However, this drawback will become an important advantage when the surface structure is becoming important, or if one would like to study three-dimensional structures.

2.1.1 Scattering from a Single Electron

X-rays are described by propagating plane waves using Maxwell's equations [17]. The wavevector $\mathbf{k}$, vector potential $\mathbf{A}$, electric field $\mathbf{E}$, and magnetic field $\mathbf{B}$ can be written as

$$\mathbf{A}(\mathbf{r}, t) = A_0 \left[a e^{i(\mathbf{k} \cdot \mathbf{r} - wt)} \boldsymbol{\varepsilon} + a^\dagger e^{-i(\mathbf{k} \cdot \mathbf{r} - wt)} \boldsymbol{\varepsilon}^* \right],$$

$$\mathbf{E}(\mathbf{r}, t) = -iwA_0 \left[a e^{i(\mathbf{k} \cdot \mathbf{r} - wt)} \boldsymbol{\varepsilon} - a^\dagger e^{-i(\mathbf{k} \cdot \mathbf{r} - wt)} \boldsymbol{\varepsilon}^* \right], \qquad (2.1)$$

$$\mathbf{B}(\mathbf{r}, t) = iA_0 \left[a e^{i(\mathbf{k} \cdot \mathbf{r} - wt)} \mathbf{k} \times \boldsymbol{\varepsilon} - a^\dagger e^{-i(\mathbf{k} \cdot \mathbf{r} - wt)} \mathbf{k} \times \boldsymbol{\varepsilon}^* \right],$$

where $\mathbf{r}$ is the real-space coordinate, A_0 is the amplitude of the vector potential, which is related to the photon energy $\hbar\omega$, $\boldsymbol{\varepsilon}$ is the incident polarization vector (electric polarisation) with $\boldsymbol{\varepsilon}^*$ its complex conjugate, and a and $a^\dagger$ are annihilation and creation operators.

Let us start by considering the most simple case of light-matter interaction in which one photon hits one electron. The non-relativistic Hamiltonian describes such a two-body system is written as $H = H_0 + H_i$, where H_0 is the non-interacting part containing the Hamiltonian of the electron (with mass m, charge $-e$, and spin $\mathbf{S}$) and the photon fields. H_i describes the photon-electron interaction [18], taking the form of:

$$H_i = \frac{e}{m} \mathbf{p} \cdot \mathbf{A} + \frac{e}{m} \mathbf{S} \cdot \mathbf{B} + \frac{e^2}{2m} \mathbf{A}^2, \qquad (2.2)$$

where $\mathbf{p}$ is the momentum operator. Let us assume that the system has an initial state $|\phi_i\rangle = |\psi_g, (\boldsymbol{\varepsilon}_i, \mathbf{k}_i)\rangle$, where $|\psi_g\rangle$ is the ground state of the electron and $|(\boldsymbol{\varepsilon}_i, \mathbf{k}_i)\rangle$ describes the incoming x-ray photon. After the interaction, the system reaches a final state $|\phi_f\rangle = |\psi_f, (\boldsymbol{\varepsilon}_f, \mathbf{k}_f)\rangle$. If the photon is eventually scattered out, $\boldsymbol{\varepsilon}_s$ and $\mathbf{k}_s$ are used, describing the scattered polarisation and wavevector.

Here, the linear combination of the three terms allows us to evaluate and separate different processes before we intend to diagonalise it (though they occur at the same time). First, the third term contains $\mathbf{A}^2$, which implies that in the elastic scattering process, the electron does nothing [19], but 'watches' the photon-in-photon-out process. This is a nonresonant scattering processes, such as Thomson scattering and nonresonant inelastic x-ray scattering. The first term contains $\mathbf{p} \cdot \mathbf{A}$, suggesting the interaction on the electric part. The second term $\mathbf{S} \cdot \mathbf{B}$ suggests the interaction between the electron spin and magnetic field component of the x-rays. Note that the annihilation/creation operator occurs only once for either $\mathbf{p} \cdot \mathbf{A}$ or $\mathbf{S} \cdot \mathbf{B}$, indicating that either a photon is absorbed, or emitted. Therefore, these two terms correspond to absorption or emission processes.

We now define a transition operator T such that $T|\phi\rangle$ excites the state of $|\phi\rangle$ into certain transition state. Moreover, $|\langle \phi_f | T | \phi_i \rangle|^2$ expects the transition probability from $|\phi_i\rangle$ to $|\phi_f\rangle$. Based on the Hamiltonian given above, T can be written as [17]:

$$T \approx T_1 + T_2 = T_1 + T_2^T + T_2^A, \quad \text{with} \qquad (2.3)$$

$$T_1 = \frac{e}{m} (\mathbf{p} \cdot \mathbf{A} + \mathbf{S} \cdot \mathbf{B}), \qquad (2.4)$$

$$T_2^T = \frac{e^2}{2m} \mathbf{A}^2 \qquad \text{and} \qquad (2.5)$$

$$T_2^A = \left(\frac{e}{m} \right)^2 (\mathbf{p} \cdot \mathbf{A} + \mathbf{S} \cdot \mathbf{B}) G_0(\mathscr{E}_i) (\mathbf{p} \cdot \mathbf{A} + \mathbf{S} \cdot \mathbf{B}), \qquad (2.6)$$

where G_0 is the Green's function of H_0, and $\mathscr{E}_i$ the initial state energy. T_1 and T_2 refer to first- and second-order processes, respectively. It is clear now that T_1 gives rise to the absorption or emission (single-photon process), while T_2 drives the scattering (two-photon process).

For a scattering experiment, the most important aspects are, (i), in what direction to expect the scattered x-rays, and (ii), what intensity to expect. The first is described by $\mathbf{q} = \mathbf{k}_s - \mathbf{k}_i$, where $\mathbf{k}_i$ ($\mathbf{k}_s$) is the wavevector of the incident (scattered) x-rays and $\mathbf{q}$ is the momentum transfer. The second is given by the double differential cross-section, $d^2\sigma/(d\Omega dw)$ [19], which is proportional to the transition probability from the initial to the final state. Usually, this can be expressed by $(d\sigma/d\Omega) \propto |f|^2$, where f is the scattering form factor.

Therefore f takes the form of

$$f = -\frac{C_s}{r_0}\langle\phi_f|T_2|\phi_i\rangle\ , \tag{2.7}$$

Note that normalisation constants of $-C_s/r_0$ are given. This is based on single free-electron Thomson scattering, which expects $f = -r_0$, and $(d\sigma/d\Omega) = r_0^2|\boldsymbol{\varepsilon}_s^* \cdot \boldsymbol{\varepsilon}_i|^2$ [1], where $r_0 = 2.82 \times 10^{-5}$ Å is the classical electron radius, or Thomson scattering length.

Equation (2.7) is the fundamental equation for calculating the transition probability for any light-scattering problem. However, it is rather trivial to solve the single free electron case as its electronic state is very simple, and the $\mathbf{P} \cdot \mathbf{A}$ (or $\mathbf{S} \cdot \mathbf{B}$) term do not play a role. This situation will drastically change by adding more electrons into the system, for example, for an atom or ion in solid states.

2.1.2 Scattering from a Single Atom

For adding more electrons to the scattering process, a phase factor $e^{i\mathbf{q}\cdot\mathbf{r}}$ has to be included in the integrated form factor for each participating electron at coordinate $\mathbf{r}$. More importantly, the electronic states (i.e., the band structure) will become the key aspect to consider.

However, we should start with the trivial case, in which only T_2^T participates. Therefore, for an atom or ion, the form factor is calculated by substituting Eqs. (2.5) into (2.7), and summing over all the occupied states:

$$f_T = \boldsymbol{\varepsilon}_s^* \cdot \boldsymbol{\varepsilon}_i \sum_g \langle\psi_g|e^{i\mathbf{q}\cdot\mathbf{r}}|\psi_g\rangle = \boldsymbol{\varepsilon}_s^* \cdot \boldsymbol{\varepsilon}_i \int \rho(\mathbf{r})e^{i\mathbf{q}\cdot\mathbf{r}}d^3\mathbf{r}\ . \tag{2.8}$$

Clearly T_2^T does nothing to the electrons, thus it is taken out of the summation/integration. This leaves only the electron density $\rho(\mathbf{r}) = \sum_g |\psi_g(\mathbf{r})|^2$ of this atom affecting the form factor. For example, in forward scattering ($\mathbf{q} = 0$), $f_T =$

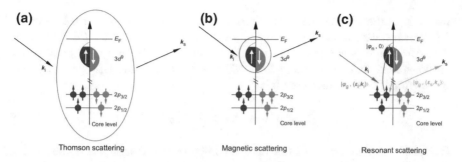

Fig. 2.1 Comparison of the three scattering processes. **a** Thomson scattering, **b** non-resonant magnetic scattering, and **c** resonant magnetic scattering off a single magnetic $3d$ ion. The sketches illustrate a Cu^{2+} ion in Cu_2OSeO_3 undergoing a $2p \rightarrow 3d$ transition

$\boldsymbol{\varepsilon}_s^* \cdot \boldsymbol{\varepsilon}_i Z$, in which Z is the atomic number. Therefore, Thomson scattering purely reflects the electron charge distribution, also called charge scattering. The physical process of Thomson scattering is illustrated in Fig. 2.1a.

The second scattering process is non-resonant magnetic scattering [9] due to the interaction between the magnetic field component of the incoming x-rays and the electron spin. Therefore the two $\mathbf{S} \cdot \mathbf{B}$ terms in Eq. (2.6) are also responsible for this interaction in the non-resonant energy regions. Now the detailed structure of $|\psi_g\rangle$ becomes important. For single-crystalline Cu_2OSeO_3, Cu is in a $2+$ state, and therefore all the occupied $|\psi_g\rangle$ states are $1s^2 2s^2 2p^6 3s^2 3p^6 3d^9$. When inserting Eqs. (2.6) into (2.7), note that the ground state of $|\psi_g\rangle$ is used on both bra and ket sides of Eq. (2.7). After summing over all the electronic states within a magnetic ion, one can see that the atomic form factor is proportional to the spin (polarised) density of states in the valence shell, as illustrated in Fig. 2.1b. However, the amplitude is typically $\sim 0.01\, r_0$, so much weaker than charge scattering [6]. In this work, we have not observed a measurable signal stemming from non-resonant magnetic scattering using both soft x-rays with photon energies well below or above the transition metal L edge.

The third scattering process is photon energy-dependent resonant scattering, also described by T_2^A [6], however, in a slightly different form. The physical process is illustrated in Fig. 2.1c, using a two-step model. First, starting from the initial state $|\phi_i\rangle = |\psi_g, (\boldsymbol{\varepsilon}_i, \mathbf{k}_i)\rangle$, the photon field excites the electronic ground state into an intermediate state $|\phi_n\rangle = |\psi_n, 0\rangle$, which then decays back to the final state of $|\phi_f\rangle = |\psi_f, (\boldsymbol{\varepsilon}_s, \mathbf{k}_s)\rangle$. The associated energies are $\mathscr{E}_i$, $\mathscr{E}_n$, and $\mathscr{E}_f$, respectively. Note that in this elastic process $\mathscr{E}_i = \mathscr{E}_f = \mathscr{E}_g + \hbar\omega$. $\mathscr{E}_n$ is the energy level corresponding to the intermediate state and $\mathscr{E}_g$ the core level energy.

The resonant scattering form factor for a single magnetic ion is obtained by inserting Eqs. (2.6) into (2.7), and iterates twice, following the process described above, and reads [20]

$$f^{\text{res}} \propto \sum_g \sum_n \frac{\langle \psi_g | \hat{O}_s^* | \psi_n \rangle \langle \psi_n | \hat{O}_i | \psi_g \rangle}{\mathscr{E}_n - \mathscr{E}_g - \hbar\omega - i\Gamma_b/2} , \qquad (2.9)$$

where Γ_b is the intermediate state lifetime, and the summation over n is over the $3d$ valence holes, while the summation over g is over all $2p$ states for the electric dipole transition. Now one can imagine that this amplitude is going to be significantly pronounced if the photon energy is tuned to be $\mathcal{E}_n = \mathcal{E}_g + \hbar w$. The operator $\hat{O}$ that comes from T_2^A takes the form of:

$$\hat{O} = \left[\mathbf{p}\cdot\boldsymbol{\varepsilon} + i\mathbf{S}\cdot(\mathbf{k}\times\boldsymbol{\varepsilon})\right] e^{i\mathbf{q}\cdot\mathbf{r}} . \tag{2.10}$$

Here, an important approximation has to be imposed that is sufficient for describing the resonant scattering process for $3d$ elements, which further decomposes $\hat{O}$ into:

$$\langle\psi_n|\hat{O}|\psi_g\rangle = i\frac{m}{h}(\mathcal{E}_n - \mathcal{E}_g)\langle\psi_n|\hat{o}_{E1} + \hat{o}_{M1} + ...|\psi_g\rangle . \tag{2.11}$$

The operator $\hat{o}_{E1}$ corresponds to the electric-dipole approximation of $\mathbf{A}$, and $\hat{o}_{M1}$ corresponds to the magnetic-dipole approximation of $\mathbf{B}$, namely

$$\hat{o}_{E1} = \boldsymbol{\varepsilon}\cdot\mathbf{p} \qquad \text{and} \tag{2.12}$$

$$\hat{o}_{M1} = \frac{\hbar}{2m(\mathcal{E}_n - \mathcal{E}_g)}(\mathbf{k}\times\boldsymbol{\varepsilon})\cdot(\mathbf{L} + 2\mathbf{S}) . \tag{2.13}$$

The $M1$ term again is small for the x-ray region, and can therefore be ignored. Consequently, $E1$ remains as the only term in transition operations of Eqs. (2.9–2.11). This is called the electric-dipole approximation for REXS.

Even so, f^{res} in Eq. (2.9) is still very large, as one has to examine all $2p$ and $3d$ states in order to write down the matrix elements. Nevertheless, one can sort out and reorganise those terms based on the angular momentum of $|\psi\rangle$. This further translates into the magnetic moment $\mathbf{m}$ of this atom. Therefore, Eq. (2.9) is rewritten, which gives the amplitude for resonant scattering at the atomic site n [2, 11]:

$$f^{\text{res}} = f_0(\boldsymbol{\varepsilon}_s^*\cdot\boldsymbol{\varepsilon}_i) - if_1(\boldsymbol{\varepsilon}_s^*\times\boldsymbol{\varepsilon}_i)\cdot\mathbf{m}_n + f_2(\boldsymbol{\varepsilon}_s^*\cdot\mathbf{m}_n)(\boldsymbol{\varepsilon}_i\cdot\mathbf{m}_n) , \tag{2.14}$$

where $\mathbf{m}_n$ is the unit magnetisation vector at the nth site, and f_0, f_1, and f_2 are energy dependent terms. Furthermore, the third term is usually very small and can be neglected. The REXS form factor for single atom can thus be approximated as:

$$f^{\text{res}} = f_0(\boldsymbol{\varepsilon}_s^*\cdot\boldsymbol{\varepsilon}_i) - if_1(\boldsymbol{\varepsilon}_s^*\times\boldsymbol{\varepsilon}_i)\cdot\mathbf{m}_n . \tag{2.15}$$

Hereby, we have ignored several terms, which do not contribute to the measurable signal at all in an experiment. First, the $\mathbf{S}\cdot\mathbf{B}$ interaction can be completely ignored, such that non-resonant magnetic scattering, do not have to be considered. Subsequently, only the $\mathbf{p}\cdot\mathbf{A}$ term matters, making the form factor summation much simpler. Second, the quadratic term in Eq. (2.14) is a higher-order effect, and can thus be omitted. After these steps, Eq. (2.15) has a linear relationship in the magnetisation vector, through which it is extremely straightforward to calculate the diffraction process as compared to neutrons [see Eq. (1.37)]. Moreover, the photon energy only adjusts the amplitude of f_0 and f_1, which only further scales the scattering intensity. This separation between the spectroscopy part (f_0, f_1) and the scattering part provides another elegant feature for the further calculations, i.e., f_0 and f_1 can be treated most of the time as constants (their actual forms are extremely complicated) for scattering events, if the photon energy does not change.

2.1.3 Scattering from a Periodic Lattice

Now we can include more atoms into the scattering process, which is nothing more than adding more phase factors $e^{i\mathbf{q}\cdot\mathbf{R}_n}$ at the atomic positions R_n into the form factor integration, assuming that the first Born approximation [2] is valid for the scattering geometry. For a periodic lattice that preserves the translational symmetry, the scattering amplitude is a delta function, giving rise to diffraction. This is true for both charge and magnetic scattering. Therefore, the basic diffraction theory that calculates the structural scattering factor within the unit cell will be applied to the magnetic crystals.

For single crystals, $\mathbf{R}_n$ is modulated in real-space and described by a certain space group. The total differential scattering cross-section is $(d\sigma/d\Omega) \propto |\mathscr{F}^{tot}|^2$, with $\mathscr{F}^{tot}$ the scattering amplitude:

$$\mathscr{F}^{tot} = \sum_{b,n} \left(f_{T(n)}^{(b)} + f_{(n)}^{res(b)} \right) e^{i\mathbf{q}\cdot\mathbf{R}_n^b} ,$$

$$= \mathscr{F}_T + \mathscr{F}_0^{res} + \mathscr{F}_1^{res} , \tag{2.16}$$

in which the summation is taken over all atomic sites n for all atom species b at the real-space positions $\mathbf{R}_n^b$. Note that, at first, non-resonant magnetic scattering is ignored. Also, the f^{res} term can be further decomposed into f_0 and f_1 terms, as shown in Eq. (2.15), resulting in $\mathscr{F}_0^{res}$ and $\mathscr{F}_1^{res}$ terms. The f_0 term essentially overlaps with the Thomson scattering term f_T, and scales with photon energy. This is the natural consequence of resonant charge scattering: for a single crystal diffraction experiment, if the x-rays wavelength is associated with a core-level transition of the specific element, the corresponding structure factor will be enhanced, giving rise to pronounced diffraction intensities for most of the structural peaks—sometimes even forbidden peaks may appear [21].

Figure 2.2 shows the basic crystallographic information for the materials investigating in detail in this work, i.e., Cu_2OSeO_3, and MnSi. The magnetic ions are Cu and Mn respectively. Both materials belong to the space group $P2_13$, in which two enantiomers will occur. Figure 2.2c and d demonstrate two MnSi unit cells with opposite chirality. Figure 2.2e–f shows the single-crystal-diffraction data for Cu_2OSeO_3, in which the hk-plane for $l = 0$ and $l = 1$ are plotted, respectively. Note that the h, k, l directions correspond to the a^*, b^*, c^* directions, where a^*, b^*, c^* are the basis vec-

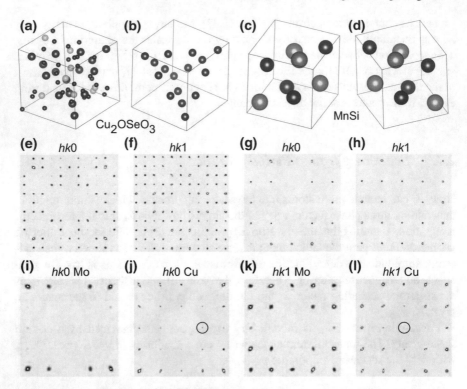

Fig. 2.2   Two typical $P2_13$ crystal structures and the corresponding single-crystal diffraction data. **a** Ball-and-stick model of the Cu_2OSeO_3 unit cell. Blue, green, and red orange balls denote Cu, Se and O atoms respectively. The positions of magnetic Cu^{2+} ions is shown and green and black balls denote Se and O atoms, respectively

tors of the reciprocal space. For a cubic system, we use the convention that h orients along $(1,0,0)$; k orients along $(0,1,0)$; and l orients along $(0,0,1)$.

A similar diffraction pattern is observed for MnSi, as shown in Fig. 2.2g–h. For such primitive cubic systems, $(0, 0, 2n + 1)$ peaks ($n = 0, 1, 2, ...$) should be forbidden for Thomson scattering as their structural factors go extinct. This is consistent with our results (see figure), when the radiation source is selected to be Mo $K\alpha$. While using Cu $K\alpha$ radiation, the results from MnSi does not change. On the other hand, for Cu_2OSeO_3, the resonant charge scattering condition is satisfied, leading to the forbidden six {001} peaks to appear, as marked by the black circle in Fig. 2.2j and l. This is explained as Templeton scattering [21], in which the anisotropic third-rank tensor stemming from the mixed dipole-quadrupole term allows for the extinction peak to appear for noncentrosymmetric crystals at or nearby the x-ray absorption edges. This is again a higher-order effect from the $\mathbf{p} \cdot \mathbf{A}$ term, and only occurs for charge scattering. Therefore we are safe to use the dipole-approximation for magnetic scattering.

It is worth noting that while performing the spatial Fourier transform based on Eq. (2.16), the most important ingredient is the reciprocal space structure, instead of the exact value of $\mathscr{F}^{\text{tot}}$. From this perspective, one can state that $\mathscr{F}_T$ and $\mathscr{F}_0^{\text{res}}$ describe almost always the same Bragg peaks (if one is not dealing with special charge structures, e.g., Templeton scattering or charge density waves), while $\mathscr{F}_0^{\text{res}}$ only enhances $\mathscr{F}^{\text{tot}}$ for that Bragg peak if it is at the resonance condition. On top of this, if this peak at $\mathbf{q}$ corresponds to the magnetic atoms, the $\mathscr{F}_1^{\text{res}}$ term can be considered. For example, for a ferromagnetic crystal at resonant condition, each Bragg peak contains the information from all three terms; for a primitive cubic antiferromagnetic crystal, besides of all the allowed Bragg peaks, an extra magnetic peak at half of the first-order lattice peak exists in reciprocal space. In summary, if the magnetic order is commensurate, all terms in Eq. (2.16) have to be looked after.

On the other hand, if the magnetic order is incommensurate such that the periodicity of the spin modulation is completely decoupled from the atomic lattice, the magnetic peaks will appear as 'satellites' that decorate each structural peak. This is the natural consequence of the Fourier transform according to Eq. (2.16). This appears to make the reciprocal space structure more complex, however, the beauty of this effect is that only $\mathscr{F}_1^{\text{res}}$ needs to be considered for each magnetic peak for a good approximation.

Concentrating only on resonant diffraction stemming from charge and magnetism, Eq. (2.16) is reduced to:

$$\mathscr{F}_0^{\text{res}}(\mathbf{q}) = f_0 \sum_n (\boldsymbol{\varepsilon}_s^* \cdot \boldsymbol{\varepsilon}_i) e^{i\mathbf{q} \cdot \mathbf{R}_n^{\text{Cu}}} = F_0 (\boldsymbol{\varepsilon}_s^* \cdot \boldsymbol{\varepsilon}_i) \overline{\rho}(\mathbf{q}) \, , \tag{2.17}$$

$$\mathscr{F}_1^{\text{res}}(\mathbf{q}) = -if_1 \sum_n (\boldsymbol{\varepsilon}_s^* \times \boldsymbol{\varepsilon}_i) \cdot \mathbf{m}_n \, e^{i\mathbf{q} \cdot \mathbf{R}_n^{\text{Cu}}} = -iF_1 (\boldsymbol{\varepsilon}_s^* \times \boldsymbol{\varepsilon}_i) \cdot \mathbf{M}(\mathbf{q}) \, , \tag{2.18}$$

where F_0 and F_1 replace f_0 and f_1 in case there are spectroscopically different species (then f_0 and f_1 also have to be Fourier transformed). In our case, although there are two crystallographically different Cu sites, they are spectroscopically identical, sharing the same Cu^{2+} valence, and carrying identical magnetic moments. Therefore, the f_0 and f_1 terms are the same for all of them. Here they describe only Cu $2p \rightarrow 3d$ electric-dipole transitions. $\mathbf{R}_n^{\text{Cu}}$ is summed over all Cu sites, giving rise to the Fourier transform of the charge density $\overline{\rho}(\mathbf{q})$ and the Fourier transform of the magnetic moment in real-space, $\mathbf{M}(\mathbf{q})$. The resonant scattering intensity is thus given by (squaring the scattering amplitude that is complex):

$$I(\mathbf{q}) = I_c(\mathbf{q}) + I_m(\mathbf{q}) + I_i(\mathbf{q}) \, , \tag{2.19}$$

where

$$I_c(\mathbf{q}) = \left| \mathscr{F}_0^{\mathrm{res}}(\mathbf{q}) \right|^2 , \tag{2.20}$$

$$I_m(\mathbf{q}) = \left| \mathscr{F}_1^{\mathrm{res}}(\mathbf{q}) \right|^2 , \tag{2.21}$$

$$I_i(\mathbf{q}) = \mathscr{F}_1^{\mathrm{res}}(\mathbf{q}) \left[\mathscr{F}_0^{\mathrm{res}}(\mathbf{q}) \right]^* - \mathscr{F}_0^{\mathrm{res}}(\mathbf{q}) \left[\mathscr{F}_1^{\mathrm{res}}(\mathbf{q}) \right]^* , \tag{2.22}$$

are the charge, magnetic, and interference terms, respectively. We define σ and π polarisation as perpendicular and parallel to the scattering plane, respectively, so that we have $\sigma_s \cdot \sigma_i = 1, \pi_s \cdot \pi_i = \hat{\mathbf{k}}_s \cdot \hat{\mathbf{k}}_i, \sigma_s \times \sigma_i = 0, \sigma_s \times \pi_i = \hat{\mathbf{k}}_i, \pi_s \times \sigma_i = -\hat{\mathbf{k}}_s$, and $\pi_s \times \pi_i = \hat{\mathbf{k}}_s \times \hat{\mathbf{k}}_i$.

For example, if σ polarisation is used for the incident beam, we can calculate the scattering intensity for an arbitrary peak $\mathbf{q}$. Following the fact that charge scattering does not rotate the light polarisation, while magnetic scattering rotates the light by $90°$, we expect σ_s for $\mathscr{F}_0^{\mathrm{res}}$, and π_s for $\mathscr{F}_1^{\mathrm{res}}$ [2]. These lead to

$$I_c^\sigma(\mathbf{q}) = \frac{1}{2} |F_0 \overline{\rho}(\mathbf{q})|^2 , \tag{2.23}$$

$$I_m^\sigma(\mathbf{q}) = \frac{1}{2} |F_1|^2 |(\hat{\mathbf{k}}_s \cdot \mathbf{M}(\mathbf{q})|^2 , \tag{2.24}$$

$$I_i^\sigma(\mathbf{q}) = 0 , \tag{2.25}$$

Note that the interference term does not vanish for π or circular polarisation.

However, if the magnetisation modulation is incommensurate, there are two consequences. First, at a diffraction condition for magnetic order ($\mathbf{q} = \mathbf{q}_m$), $I_c(\mathbf{q}_m) = 0$, implying that any charge properties due to the crystallography will not interrupt the magnetic diffraction process. Second, because of this, no interference occurs at those incommensurate magnetic peaks, meaning that $I_i(\mathbf{q}_m) = 0$. Eventually, if we only intend to study the magnetic peaks, Eq. (2.18) represents the only term to be considered. Moreover, F_1 does not play a role for diffraction, leaving only the light polarisation and the magnetic structure playing a role in the end.

At this point, to get a first impression of resonant magnetic diffraction patterns of magnetic orders in helimagnets, no knowledge of the REXS process (that we have just written down) is required at all. The diffraction pattern is nothing but $\mathbf{M}(\mathbf{q})$, i.e., the Fourier transform of the real-space magnetisation.

2.1.4   Scattering from Helical, Conical, and Skyrmion Order

We will now plug in some real magnetic structures observed in Cu_2OSeO_3, and calculate the scattering form factors. These results can be directly used to compare with the experimental data. The magnetisation modulation has a long wavelength of $\lambda_h \approx 60$ nm [22], which is the number that we will use for the calculation. Note that in reciprocal space lattice unit, this value corresponds to 0.0149 (r.l.u.).

The magnetic properties of Cu_2OSeO_3 are well described by the universal heli-magnetic theory introduced in Chap. 1. As the continuum approximation is valid, the discrete magnetic moments $\mathbf{m}_n$ associated with the magnetic ions in Eq. (2.18) are approximated by a continuous vector field of the magnetisation, $\mathbf{m}(\mathbf{r})$. Referring to Eqs. (1.6)–(1.8), the energy density can be written as:

$$w(\mathbf{m}) = A (\nabla \mathbf{m})^2 + D\mathbf{m} \cdot (\nabla \times \mathbf{m}) - \mathbf{m} \cdot \mathbf{B} + w_A \ , \qquad (2.26)$$

where w_A the anisotropy term.

The system's ground state ($\mathbf{B} = 0$) is the one-dimensional, helically ordered state composed of single-harmonic modes. It takes $\xi = 90°$ in Eq. (1.15), written as:

$$\mathbf{m}(\mathbf{r}) = M_S[\mathbf{n}_3\cos(\mathbf{q}_h \cdot \mathbf{r}) + \mathbf{n}_2\sin(\mathbf{q}_h \cdot \mathbf{r})] \ , \qquad (2.27)$$

with the magnetisation vector $\mathbf{m}(\mathbf{r}) = m_1(\mathbf{r})\mathbf{n}_1 + m_2(\mathbf{r})\mathbf{n}_2 + m_3(\mathbf{r})\mathbf{n}_3$. $\mathbf{q}_h$ is the wavevector of the helix, with $\lambda_h = 2\pi/q_h$ being the real-space helical pitch. The orientation of the modulation at $\mathbf{B} = 0$ is locked along a $\langle 100 \rangle$ direction by the cubic anisotropy [23]. The two-dimensional magnetisation configuration for one helical pitch is illustrated in Fig. 2.3a, for which the modulation is along x. This is the motif of the magnetic crystal. The form factor of a helix is, following Eq. (2.18), given by

$$f_h = \mathbf{V} \int_0^{\lambda_h} (\boldsymbol{\varepsilon}_s^* \times \boldsymbol{\varepsilon}_i) \cdot [\mathbf{n}_3\cos(\mathbf{q}_h \cdot \mathbf{r}) + \mathbf{n}_2\sin(\mathbf{q}_h \cdot \mathbf{r})]e^{i\mathbf{q}\cdot\mathbf{r}}dr \ , \qquad (2.28)$$

with $\mathbf{V} = -iM_Sf_1$ being a constant. Note that an integration over the entire helical pitch replaces the summation in the continuum approximation. Such long-range-ordered helical crystal has a 'unit cell' that consists of two helix motifs, separated by a distance λ_h, as shown in Fig. 2.3c, with the structural form factor

$$F_h = f_h \left(1 + e^{i\mathbf{q}\cdot\lambda_h\hat{\mathbf{R}}}\right) \ . \qquad (2.29)$$

Using σ-polarised incident light, the calculated $|f_h|^2$ is plotted in Fig. 2.3b, which shows the REXS form factor of the helix motif. The structure factor, which essentially demonstrates the diffraction pattern, is calculated and shown in Fig. 2.3d.

Above a certain magnetic field B_{c1}, the conical spiral state forms with a finite ξ, taking the form of Eq. (1.15). The wavevector $\mathbf{q}_h$ is along the magnetic field. At a certain field B_{c2}, all the magnetisation vectors are parallel, forming the ferrimagnetic state.

The conical periodic order can also be described by a unit cell that consists of two conical spirals with the 'lattice constant' of λ_h. As for the helical state, the resonant scattering structure factor can be written as

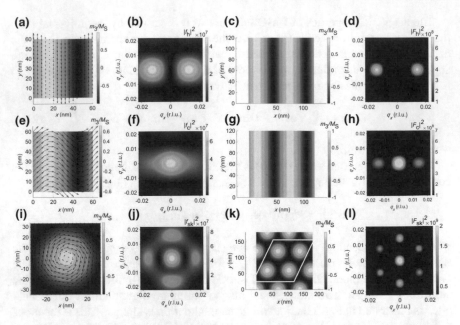

Fig. 2.3 REXS structure factor calculations for various magnetic orders in $P2_13$ helimagnets. **a** Magnetisation motif for the helical order. **b** REXS form factor $|f_h|^2$ for the helical motif, using σ-polarised incident x-rays. **c** Helical lattice unit cell, consisting two helical motifs. **d** Calculated diffraction pattern, using σ-polarised incident light. **e–h** The same calculations for the conical magnetic order. **i–l** Calculations for the skyrmion lattice state. Note that the unit cell is marked by the white lines in (**k**)

$$F_c = f_c \left(1 + e^{i\mathbf{q}\cdot\lambda_h\hat{\mathbf{R}}}\right) , \quad \text{with}$$

$$f_c = \mathbf{V} \int_0^{\lambda_h} (\boldsymbol{\varepsilon}_s^* \times \boldsymbol{\varepsilon}_i) \cdot \left[\cos\xi\mathbf{n}_1 + \sin\xi\cos(\mathbf{q}_h \cdot \mathbf{r})\mathbf{n}_2 + \sin\xi\sin(\mathbf{q}_h \cdot \mathbf{r})\mathbf{n}_3\right] e^{i\mathbf{q}\cdot\mathbf{r}} dr .$$

$$(2.30)$$

The calculated results are shown in Fig. 2.3e–h.

A single skyrmion structure is specified by Eq. (1.11), in which the three magnetisation components $(m_1^{\text{sk}}, m_2^{\text{sk}}, m_3^{\text{sk}})$ are given. The form factor for a individual skyrmion can be written in the form of

$$f_{\text{sk}} = \mathbf{V} \iint_{\text{sk}} (\boldsymbol{\varepsilon}_s^* \times \boldsymbol{\varepsilon}_i)(m_1^{\text{sk}}\mathbf{n}_1 + m_2^{\text{sk}}\mathbf{n}_2 + m_3^{\text{sk}}\mathbf{n}_3)e^{i\mathbf{q}\cdot\mathbf{r}} d\mathbf{r}. \quad (2.31)$$

The integral is carried out over the circular area of a skyrmion vortex. The 'crystalline' structure is a hexagonal-type two-dimensional lattice. Therefore, a hexagonal two-dimensional unit cell can be chosen, as indicated by the white lines in Fig. 2.3k. The structure factor then becomes

$$F_{\text{sk}} = f_{\text{sk}}(1 + e^{i\mathbf{q}\cdot\mathbf{a}_1} + e^{i\mathbf{q}\cdot\mathbf{a}_2} + e^{i\mathbf{q}\cdot\mathbf{a}_3}) , \quad (2.32)$$

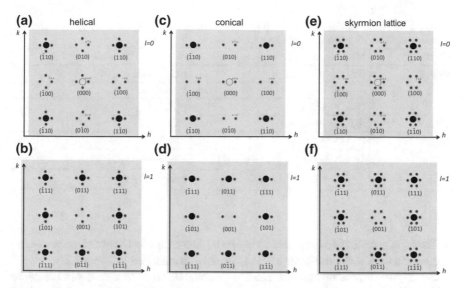

Fig. 2.4 Reciprocal space structure of the magnetic orders in Cu_2OSeO_3 as seen by REXS. Two reciprocal space planes are shown: $hk0$ and $hk1$, respectively. The black dots are the charge peaks, while the red dots are the magnetic peaks

where $\mathbf{a}_1, \mathbf{a}_2, \mathbf{a}_3$ are the real-space base vectors, which are rotated by $60°$ with respect to each other. The core-to-core distance is $a_1 = a_2 = a_3 = 2\lambda_h/\sqrt{3}$, which can be regarded as the 'lattice constant' of the skyrmion crystal. The calculated form factor, as well as the diffraction pattern, is shown in Fig. 2.3i–l.

The last step is to assemble the magnetic peaks and the structural peaks into one picture. As discussed before, this is simply achieved by imposing the incommensurate magnetic peaks onto each structural lattice peak. We would like to emphasise that the lattice peak here means the reciprocal space lattice points—even if some of them does not have Thomson structure factor. In other words, the magnetic peaks should also decorate the crystallographically forbidden peaks. Again, this is the natural consequence of $\mathbf{M}(\mathbf{q})$. Therefore, the magnetic peaks actually exist at $\mathbf{Q} = \mathbf{G} + \mathbf{q}_m$, where $\mathbf{G}$ is a lattice peak, and $\mathbf{q}_m$ is the magnetisation modulation wavevector. The large-area reciprocal space structure is demonstrated in Fig. 2.4, using the parameters for Cu_2OSeO_3.

For Cu_2OSeO_3, the $(0,0,1)$ Bragg peak is the only accessible structural peak for the x-ray wavelength corresponding to the Cu L_3 resonance. However, it is sufficient to find all magnetic peaks surrounding it. Taking cubic anisotropy into account, three helical domains that propagate along the three $\langle 001 \rangle$ directions are to be expected. For the conical phase, only single conical domain that is driven along the magnetic field direction is to be expected. For the skyrmion lattice phase, six-fold-symmetric magnetic peaks will be present, with one pair of the three locked along $\langle 001 \rangle$ direction, if the geometry is allowed. Therefore, if we have a [001]-oriented single crystal,

Table 2.1 Magnetic modulation vectors for the magnetic phases of Cu_2OSeO_3, and the associated magnetic satellites observable in a REXS experiment

Phase	Modulation vectors	Magnetic reflections
Helical	$(0,0,q_h), (0,0,-q_h)$	$(0,0,1 \pm q_h)$
	$(q_h,0,0), (-q_h,0,0)$	$(\pm q_h,0,1)$
	$(0,q_h,0), (0,-q_h,0)$	$(0,\pm q_h,1)$
Conical	$(0,0,q_h), (0,0,-q_h)$	$(0,0,1 \pm q_h)$
Skyrmion	$(\tau,0,0), (-\tau,0,0)$	$(\pm\tau,0,1)$
	$\left(-\frac{1}{2}\tau, \frac{\sqrt{3}}{2}\tau, 0\right), \left(\frac{1}{2}\tau, -\frac{\sqrt{3}}{2}\tau, 0\right)$	$\left(\mp\frac{1}{2}\tau, \pm\frac{\sqrt{3}}{2}\tau, 1\right)$
	$\left(-\frac{1}{2}\tau, -\frac{\sqrt{3}}{2}\tau, 0\right), \left(\frac{1}{2}\tau, \frac{\sqrt{3}}{2}\tau, 0\right)$	$\left(\pm\frac{1}{2}\tau, \pm\frac{\sqrt{3}}{2}\tau, 1\right)$

while the magnetic field is applied along [001], the reciprocal space structure of the magnetic orders that surrounding (0,0,1) peak is summarised in Table 2.1.

So far, we have addressed the most fundamental issue for a resonant magnetic diffraction experiment, i.e., how the magnetic peaks look like. These are summarised in Figs. 2.3 and 2.4, which provide the characteristic diffraction patterns that can be used for distinguishing the helical, conical and skyrmion lattice states. To this point, we only used one polarisation of the incident light (σ), and Eq. (2.24) is valid for all cases. This is sufficient for the magnetic-phase-distinguishing purposes, which turns out to be the primary information to be obtained when studying skyrmion-carrying systems. Let us keep using σ-polarisation for a while to obtain more structural information about the long-range magnetic orders. Starting in Chap. 4, we will explore the effects of light polarisation, which elevates our understanding of skyrmions to the next level.

2.2 REXS in Reflection Geometry

For Cu_2OSeO_3, the (0,0,1) peak can be easily reached using REXS in reflection geometry. The measurements were carried out on beamline I10 at the Diamond Light Source, in the ultrahigh vacuum diffractometer RASOR [24]. Figure 2.5a illustrates the geometry of the experiment. The σ-polarised soft x-rays are tuned to the Cu L_3 edge at a photon energy of 931.25 eV. The scattered light is captured by either a CCD camera or a photodiode at an angle 2Θ with respect to the incoming beam. The adjustable beam size, which defines the sampling area of REXS, can vary from $20 \times 20\ \mu m^2$ to $1 \times 1\ mm^2$. The exposure time is usually kept at 2 ms for a single image, though it is also adjustable. A carefully polished high-quality Cu_2OSeO_3 single-crystal, measuring $5 \times 3 \times 0.5\ mm^3$, was positioned as sketched with the edges along [110], [1$\bar{1}$0], and [001]. The sample was confirmed to be of high quality by means of single-crystal x-ray diffraction on a Rigaku SuperNova using both Mo $K_{\alpha 1}$ and Cu

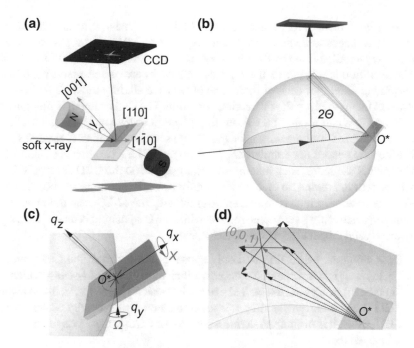

Fig. 2.5 Experimental REXS setup. **a** Scattering geometry and field configuration in RASOR. The magnetic field is provided by permanent magnets in a rotatable variable field assembly. **b–d** Ewald sphere representation of REXS; **c** and **d** are the zoomed-in illustrations of (**b**). In (**c**), the standard four-circle diffractometer is presented in reciprocal space. The goniometer is driven by three axes, in which Ω is the major axis rotating within the scattering plane. The $\mathscr{X}$ axis is coupled to Ω, while the Ψ axis is coupled to $\mathscr{X}$. This allows the three rotation matrices to be commutative. In (**d**), the (0,0,1) structure peak is marked as the blue arrow, while the magnetisation propagation wavevectors for the skyrmion lattice phase is labelled by the six red arrows. This gives rise to the six magnetic peaks in reciprocal space, as labelled by black arrows

$K_{\alpha 1}$ sources, see Fig. 2.2. It is of uniform lattice chirality as determined by electron backscatter diffraction (EBSD) [25].

For measurements on RASOR, the magnetic field direction is typically fixed with respect to the sample plane, i.e., both the magnets and the sample are moved together with the goniometer angle Ω. However, the magnets can also be rotated away from the surface normal direction in the scattering plane by a tilt angle γ, where $\gamma=0°$ corresponds to the field along the [001] axis of the sample.

The goal of the experiments is to map out the intensity distribution spanning the reciprocal space. This activity is called reciprocal space mapping (RSM). In the Ewald representation, as shown in Fig. 2.5b, in order to bring a reciprocal space point **Q** into the diffraction condition, one has to rotate the sample such that **Q** 'touches' the surface of the sphere. This rotation usually involves three axes, Ω, $\mathscr{X}$ and Ψ, as shown in Fig. 2.5c. If a point detector is used, one not only has to move **Q** onto the sphere, but also bring it into the scattering plane. Therefore, at least two circles have to be implemented for an arbitrary $\mathbf{Q}$. On the other hand, if an area detector is used, only Ω needs to be adjusted, because $q_h = \tau \approx 0.0158$ (r.l.u.) is very small. Thus, a very small volume of the reciprocal space has to be mapped. For example, for the skyrmion lattice phase, the six magnetic peaks are represented by the black arrows in Fig. 2.5d. For each of them, one only has to slightly move Ω (differently for each of them) to reach the diffraction condition. The CCD camera is mounted on the 2Θ-arm that can rotate in the scattering plane in order to capture the scattered x-rays. Each set of goniometer angles $(\Omega, \mathcal{X}, \Psi)$ selects a particular range of $\mathbf{Q}$.

Further, we choose the coupled Ω–2Θ scan mode, i.e., 2Θ rotates twice as fast as Ω for the RSM scans, such that the specular position on the CCD camera is fixed. It is important to emphasise that for RSM only a small volume in reciprocal space needs to be covered, which is equivalent to employing the Ω-2Θ scan mode or the Ω scan mode (where only the Ω stage rotates, while the 2Θ position is fixed). However, there are several advantages of the Ω-2Θ method:

(i) RSM along the l-direction will only appear in the centre of the CCD camera. Therefore it is more accurate and very efficient to carry out l-scans, which are important for mapping the helical and conical states. Moreover, as the structural $(0,0,1)$ peak is extremely strong, leading to blooming especially of the vertical CCD pixels, it is difficult to define a peak in the CCD camera image, unless 2Θ is coupled to Ω.

(ii) The adding-up of a full series of Ω–2Θ RSM scan images yields the symmetries of the the magnetic satellites and their positions relative to the $(0,0,1)$ Bragg peak. This integrated image is accurate enough to discriminate between helical, conical, skyrmion, ferrimagnetic, and paramagnetic orders, thus allowing for an efficient mapping of the phase diagram. However, quantitative data processing that projects the camera information into reciprocal space has to be performed.

(iii) The number of events for a magnetic satellite in an integrated image is equal to the integrated intensity under the 'rocking curve' of that magnetic peak. This provides more accurate data for further quantitative analysis of the REXS intensity.

2.2.1 Results

Cu_2OSeO_3 has a uniform magnetic phase diagram that is independent of the field-cooling history. Starting from a zero-field-cooled system just below the ordering temperature of 57 K, the helical state should be stabilised, giving rise to three differently oriented helical domains. As shown by the orange arrows in Fig. 2.6a, they should be locked along h, k and l, respectively. Therefore, if a Ω-2Θ-type RSM scan is carried out, the four wavevectors that lie on the hk-plane ($l = 1$) should be visible in the camera image, while the two peaks along l should always be on the specular position of the CCD. The integrated RSM CCD image is shown in Fig. 2.6b. The beam size used here is 0.5×0.5 mm^2. The diffractometer is aligned with the $(0,0,1)$ peak, thus the brightest spot in the centre corresponds to the resonant $(0,0,1)$

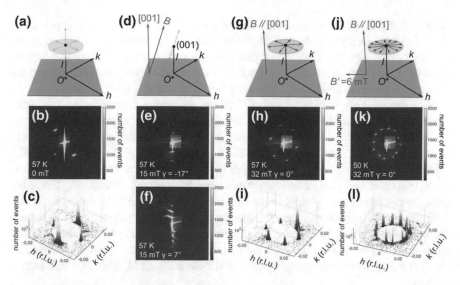

Fig. 2.6 REXS results revealing the long-range magnetic orders of the helical, conical and skyrmion lattice phase. **a, d, g, j** Three-dimensional reciprocal space structures of the REXS intensity for helical, conical, single-domain skyrmion lattice and metastable skyrmion lattice states. **b, e, h, k** Integrated RSM images for each phase. **e, f, i, l** Processed data plotted in the $hk1$ plane, in which magnetic satellites are observed

charge peak. The violation of the Bragg extinction condition is most likely due to the Templeton scattering effect, as discussed before. The diffuse intensity on the camera (along the vertical direction) is due to the flooding effect of the overexposure in the centre. If zoomed in, a square shape that is weaker than the flooded line can be observed in the centre of the image. This is the specular reflection from the sample as a whole: the resonant scattering from the charge is so strong that the fluorescence illuminates the entire sample (or an even larger area), and projects this real-space image onto the camera. These features are thus the artefacts.

The helical magnetic satellites, shown as the four spots, can be clearly identified. They correspond to two different helical domains, and each of them gives rise to a Friedel pair (i.e., $+\mathbf{q}_h$ and $-\mathbf{q}_h$ wavevectors). Detailed data processing, that projects the RSM scan images into reciprocal space $hk1$ plane, results in Fig. 2.6c, which is in good agreement with the helimagnetic theory, as well as the REXS theory predicts (see last section). The measured $q_h \approx 0.0158$ (r.l.u.) corresponds to a helical pitch of 55–56.5 nm in real space. Note that the brightness of the diffraction from one pair (locked along k) is stronger than the other (locked along h), suggesting that one domain size is larger than the other, in the area probed by the beam.

By increasing the magnetic field slightly to the critical field B_{c1}, the helical domains form a single-domain conical state. The conical propagation wavevector $\mathbf{q}_h$ is along the field direction. Therefore, if the magnetic field is along [001], the magnetic satellites will lie along l, which is not visible in the camera. In order to show the Friedel pair of the conical peaks on the CCD image, a slight rotation of the field angle by γ has to be performed. As shown in Fig. 2.6d, if the conical pair deviates from l, the vertically aligned satellites will show up on the area detector, as can be seen in Fig. 2.6e and f, for which different γ angles were used. Note that higher-order peaks are also observable. By performing data processing, the real-space pitch of the conical modulation is essentially identical to that of the helical state.

By increasing the field at 57 K to 32 mT, i.e., above the critical field B_{A1}, a first-order transition from the conical state to the skyrmion lattice state will take place. This will manifest itself as a sudden change from two- to six-fold-symmetric peaks in reciprocal space, as shown in Fig. 2.6j. The skyrmion lattice plane contains the [100] crystallographic orientation, thus one pair of the three skyrmion satellites is locked along that direction. The RSM integrated image for this state is shown in Fig. 2.6h, with the processed data plotted in the $hk1$ plane as shown in i. As observed in Fig. 2.6h, the six diffraction spots on the camera do not form a perfect circle, however, show an elliptical shape centred at $(0,0,1)$ peak. This is a purely geometrical effect at an angle of $2\Theta \approx 96.5°$. After processing the camera images in reciprocal space, the six magnetic peaks form a perfect circle, as shown in Fig. 2.6l, and in the subsequent figures.

Further, the field-polarised state is reached when the magnetic field is increased along [001], leading to the disappearance of the magnetic satellites. Another state is the paramagnetic phase, in which no magnetic order exist at all. Therefore, by measuring the diffraction pattern using σ-polarised light, all ordered magnetic states can be unambiguously resolved.

Furthermore, another phenomenon that has not been observed before was discovered in our experiment, the metastable skyrmion lattice state. It is obtained by imposing an extra biasing in-plane field. A metastable skyrmion lattice phase had been realised before in two ways. First, in $Fe_xCo_{1-x}Si$, the skyrmion lattice can survive at very low temperatures if the system is field-cooled from the paramagnetic state [26, 27]. The required field during cool-down has to be the field that stabilises a normal skyrmion lattice in a zero-field-cooled scenario. However, in the entire $P2_13$ family, $Fe_{1-x}Co_xSi$ is the only material showing this behaviour. Second, the metastability can be obtained by quenching from the stable skyrmion phase, as demonstrated for MnSi [28].

For Cu_2OSeO_3, we first stabilise the standard skyrmion phase by applying a 32 mT field along [001] at 57 K. Then, a secondary, 6 mT magnetic field is applied along [110] (provided by a Helmholtz coil pair). We can now observe twelve-fold magnetic peaks, corresponding to two domains of the skyrmion lattice state, as shown in Fig. 2.6j–l. This is astonishing as one would expect that the skyrmion tubes are tilted by the total vector field, leading to a titled skyrmion lattice plane. As shown in Fig. 2.6l, the $hk1$ plane still contains the magnetic peaks that form a perfect circle, suggesting that the skyrmion lattice plane does not tilt with the magnetic field vector. Moreover, due to the sample geometry, if the skyrmion lattice forms in the (001) plane, there are two equivalent axes, [100] and [010], that are available for one of the Friedel pairs of the skyrmion peaks to lock on to. Due to this degeneracy, two equivalent domains are observed. This resembles the metastability observed in $Fe_{1-x}Co_xSi$ [29], in which two equivalent skyrmion lattice domains are found in neutron diffraction experiments,

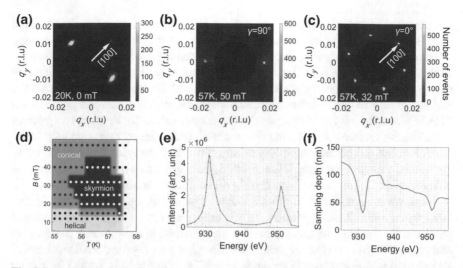

Fig. 2.7 REXS data for a Cu_2OSeO_3 sample with opposite crystalline chirality. **a–c** Reciprocal space map in the $hk1$ plane for helical, conical ($\gamma = 90°$), and skyrmion lattice phases, respectively. **d** Magnetic phase diagram mapped by REXS, using $\gamma = 0°$, field-cool protocol. Reprinted from Ref. [30]. Copyright 2016 by American Chemical Society. **e** Photon energy dependence of the total magnetic satellites intensity. **f** Calculated penetration depth as a function of photon energy for our scattering geometry

manifesting itself as 12 diffraction spots. More importantly, when cooling down, the system in such a field setup can support the skyrmion lattice phase down to 20 K, which is about 35 K lower than the skyrmion-to-conical phase transition temperature. This confirms the metastability of the state. Such a state does not decay on the time scale of one hour, which is the longest timescale we can afford to wait during an experiment at the synchrotron.

Note that this phenomenon has not been observed in Cu_2OSeO_3 before using other magnetic characterisation techniques. This highlights one of the unique properties probed by REXS: the surface. Next, we will give a short interpretation of the microscopic structure. From a bulk property point of view, the skyrmion tubes will follow the field direction. As discussed, the perpendicular main magnetic field and an in-plane bias field give rise to a titled field, as if γ was rotated. This is the case when the bias field is above certain threshold value, i.e., ∼6.5 mT. Below the threshold value, the magnetic satellites form a plane that is always perpendicular to the main field. The penetration depth profile as a function of photon energy at an incident angle of ∼48.2° is shown in Fig. 2.7f, calculated by Gerrit van der Laan from Diamond Light Source [31]. At the Cu L_3 edge, the probing depth is only 34 nm, meaning that we probe the top surface only. Therefore, below the threshold value, the bulk skyrmion tubes (deep in the sample) tilt along the vector field. For the surface magnetic structure, the skyrmion lattice plane tends to align with the surface, owing to the boundary conditions, imposed by the shape of the sample. Note that this has nothing to do with the demagnetisation field, which is also a shape effect. In our case, this abrupt change from bulk to boundary requires the spin modulations to lie in the plane that is defined by the surface boundary, in order to lower down the energy. This introduces a different behaviour of the bulk and the surface magnetic structure. However, breaking the skyrmion tube into discontinuous 'fractures' will cost much more energy. Consequently, the skyrmion tubes will gradually 'bend' towards the surface, giving rise to a different surface magnetic structure that we probed with REXS. When the bias field is larger than threshold value, the Zeeman energy term overrides the surface effect, giving rise to a uniformly aligned skyrmion tube structure. Note that this phenomenon only occurs in the field-applying protocol known to introduced a metastable state. If the system is field-cooled down from paramagnetic phase with a tilt γ angle, the common skyrmion lattice features will recover.

Next, we used another Cu_2OSeO_3 single crystal with opposite crystalline chirality in the same REXS geometry. In this case, the spin spirals in the helical and conical states, as well as the skyrmion vortex, will have the opposite handedness compared with the previous crystal. The diffraction patterns for the helical, conical and skyrmion lattice phases is shown in Fig. 2.7a–c (in the $hk1$ plane). Clearly, no difference is observed using σ-polarised light: the magnetic peaks, the wavevectors, and the propagation locking orientations are identical with the results obtained for the other chirality. Therefore, linearly polarised light is not sensitive to chirality.

The phase diagram mapped out by REXS is shown in Fig. 2.7d, which is consistent with the magnetometry data reported so far [22, 32]. However, as each image only takes 2 ms to record, the data acquisition time for a phase diagram is less than an hour, being another advantage of this technique. Figure 2.7e shows the energy dependence of the magnetic peak intensities. The absorption spectrum profile suggests that the diffraction has a magnetic origin, confirming that the theories we have been using or developing so far are correct.

In summary, reflection REXS shows the magnetic peaks are located around the $(0,0,1)$ structural peak, and the diffraction patterns show stark contrasts allowing to distinguish between the different magnetic phases. Second, this technique is sensitive to all three components q_x, q_y and q_z, of a magnetic wavevector $\mathbf{q}_m$ (while neutrons are only sensitive to q_y and q_z components). This allows us to obtain the complete set of information about the magnetic structure. e.g., for the conical order, as well as the titling of the skyrmion plane, etc., which SANS and LTEM unable to provide. The probing area can vary from mm to few tens of μm, and the measurement time is very short in comparison. Third, reflection REXS is surface sensitive, therefore detailed information about the technologically relevant surface state can be obtained.

2.3 REXS in Grazing Geometry

We now turn to a more general question: how can one deal with other B20 skyrmion-carrying compounds using soft x-rays? Apparently the $(0,0,1)$ peak is not accessible for MnSi, FeGe, $Fe_{1-x}Co_xSi$, MnGe, and so on. Nevertheless, Fig. 2.4 suggests that even around the reciprocal space origin $(0,0,0)$, there are magnetic satellites which

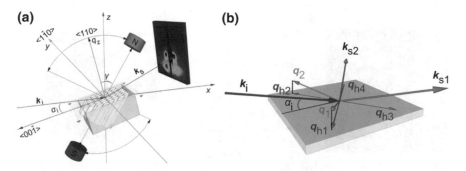

Fig. 2.8 Grazing incidence REXS geometry and setup. **a** Scattering geometry, showing the sample orientation and the magnetic field setup. The σ-polarised incident beam is tuned to the Mn L_3 edge with α_i varying from $0°$ to $10°$. **b** The magnetic peaks that are parallel to the q_x-q_y plane are labelled by the blue vectors. At low incident angles that are just above the critical angle, the reflected beam can pick up the diffraction signal that comes from the surface modulations. This is explained by the surface diffraction with truncation rods. Therefore, the propagation wavevectors 'transfer' into the 'pulled-up' scattering wavevectors, as labelled by the orange vectors. They lead to magnetic diffraction, with the diffracted light labelled by the green vectors

should be measurable. This is the basis for the transmission magnetic diffraction geometry, for both neutrons [33] and x-rays [34].

The other possibility for mapping out the reciprocal space around (0,0,0) is to lower down the diffraction angle in the grazing geometry. This is perfectly possible from a theoretical perspective [35, 36], however, may become extremely difficult in an experiment. At lower angles, many optical effects may occur at the same scattering angle. This will either overshadow the magnetic features, or at least increase the possibility of data misinterpretations. For example, if the grazing-incident geometry is used, the reflectivity will dominate the signal, including the strong specular, surface roughness correlations and any structural defects on the surface [37]. As a result, more efforts need to be devoted to clarify the origin of the scattering signals.

This geometry was used to characterise the magnetic order in MnSi single crystals. The MnSi bulk sample quality was checked by single-crystal diffraction using both Cu and Mo radiation sources (see Fig. 2.2). A [110]-oriented sample with a carefully polished surface was mounted in the soft x-ray diffractometer, as sketched in Fig. 2.8a. An external magnetic field is also provided by permanent magnets (as in the Cu_2OSeO_3 study). This sample orientation allows for most of the magnetic propagation wavevectors lying in the x-y-plane. For example, in the ground state, two pairs of helical peaks are locked along $\langle 111 \rangle$, which are within the x-y-plane. By applying a finite magnetic field with $\gamma = 0°$, the skyrmion lattice plane is also formed within the x-y-plane, while one pair out of the six magnetic peaks is locked along $\langle 110 \rangle$ directions (for **B** is not along $\langle 100 \rangle$ directions)[33] that are accessible in this geometry.

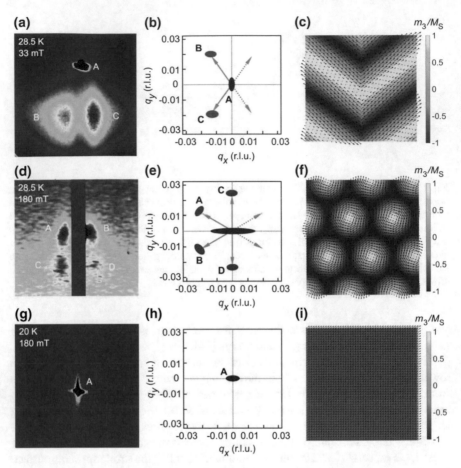

Fig. 2.9 Grazing incidence REXS results on a MnSi single crystal. The incident beam is tuned to be at the Mn L_3 edge, i.e., 643 eV with σ-polarisation. The incident angle α_i varies from $0°$ to $10°$. **a–c** Grazing REXS pattern and the reciprocal space correspondence of the labelled peaks in the helical state, as well as the real-space magnetisation texture on the surface of the sample; **d–f** skyrmion lattice state, and **g–i** conical state

The reason why resonant x-rays can pick up the magnetic peaks that completely lie on the surface is due to the surface diffraction principle [35], which is illustrated in Fig. 2.8b. When the incident angle α_i is just above the critical angle, the x-ray waves become extremely surface sensitive. For example, in the helical state, two differently oriented helical domains give rise to four helical peaks q_{h1}–q_{h4}, as labelled by the blue vectors in Fig. 2.8b. In the surface diffraction scenario, truncation rods [35] extending up along the q_z direction form. In other words, the magnetic peaks 'transfer' into rods that are 'pulled up' from the two-dimensional reciprocal space. This allows the effective scattering wavevector q_1 and q_2 to give rise to magnetic diffraction. On the other hand, at such small α_i angles, there are no possible diffraction conditions for q_{h3} and q_{h4}, even if the truncation rods extend up, to obtain the q_z component.

The experimental proof of resonant magnetic surface diffraction is demonstrated in Fig. 2.9a. MnSi has a surface critical angle of $\theta_c \approx 1.5°$. At 28.5 K and 33 mT with $\gamma = 0°$, it is in the helical state. This gives rise to two visible surface diffraction peaks B and C under $\alpha_i = 2.3°$, directly captured by the CCD camera. Spot A is the specular reflectivity, while q_{h3} and q_{h4} are not available in this geometry. The scattering wavevectors q_1 and q_2 can be directly measured. By projecting them onto the $(hk0)$ reciprocal space plane, the modulation wavevector can be unambiguously determined to be $q_h = 0.026 \pm 0.002$ (r.l.u.), see Fig. 2.9b, This corresponds to a real-space helical pitch of $\sim$17.8 nm, in agreement with both SANS and LTEM results [33, 38].

Increasing the magnetic field to 180 mT will induce the formation of the skyrmion lattice phase. This gives rise to the diffraction pattern shown in Fig. 2.9d. Projecting the camera image back into reciprocal space results in four magnetic peaks A, B, C and D, which are separated by about 60°. The length of the magnetic wavevector is same as the helical wavevector. Further decreasing the temperature will drive the system into conical phase, with the conical propagation wavevector perpendicular to the sample surface. The surface magnetic structure for the conical state is shown in Fig. 2.9i. Therefore, no surface diffraction is expected. This indeed leads to the vanishing of the magnetic peaks, leaving only the specular spot A.

In summary, by employing grazing-incidence REXS, the magnetic long-range order on the surface level can be measured. The technique is extremely sensitive to the magnetic peaks that lie in the $(hk0)$ reciprocal space plane. However, only the negative q_x part is visible in the surface diffraction conditions with low α_i. Increasing α_i, on the other hand, will eventually satisfy the diffraction conditions for the positive q_x parts. Nevertheless, the surface sensitivity may reduce drastically when increasing the incidence angle, leaving no measurable signal in the end. Though being limited by the small incident angle, under this geometry, it is possible to unambiguously distinguish between the helical, conical and skyrmion lattice phases. Consequently, grazing REXS is a universal technique that can be used to study all $P2_13$ helimagnets.

Moreover, if the magnetic peak is not within the q_x–q_y plane (like the conical peaks shown above), it will not give rise to surface diffraction at an arbitrary α_i angle. Nevertheless, this magnetic peak can be effectively measured using the reflection REXS geometry at $(0, 0, q_h)$, though the diffraction angle is also very small.

To best illustrate this, we will turn to another B20 material, FeGe thin films. B20-phase FeGe thin films were grown by magnetron sputtering, with the film thickness ranging from 40–200 nm. Unlike previous work on sputtered and molecular beam epitaxy-grown MnSi [39–41] and FeGe [42–44] thin films, which was performed on (reactive) Si(111) substrates, we use inert, (0,0,1)-oriented MgO. This brings about two major advantages: (i) the films can be deposited directly onto the substrate without the need for a seed layer, largely eliminating complications arising from the formation of a magnetically active interfacial layer [41]; (ii) the lattice mismatch between MgO ($a = 4.13$ Å) and FeGe ($a = 4.70$ Å) leads to compressive instead of tensile strain in the FeGe films, which results in an enhancement of T_c. For our

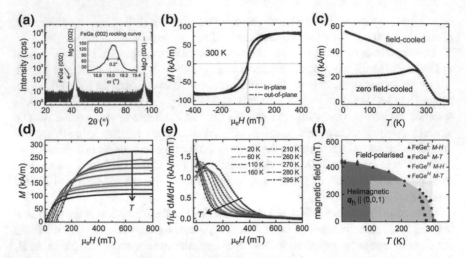

Fig. 2.10 Structural and magnetic study of 206 nm-thick FeGe films. **a** Out-of-plane XRD showing a preferred (002) orientation of the FeGe film. **b** Room-temperature magnetisation measurement with the field applied both in-plane and out-of-plane. The temperature dependence of the magnetisation is shown in (**c**). The zero field-cooled (ZFC) curve is obtained by first cooling the sample in zero field from 300 K down to 10 K before measuring in an out-of-plane field of 20 mT while heating. The field-cooled (FC) data are obtained by first cooling in an applied field of 500 mT again down to 10 K, before measuring in an out-of-plane field of 20 mT while heating. **d** Magnetisation as a function of out-of-plane field at different temperatures. The legend is shown in (**e**). **e** Susceptibility as a function of field at various temperatures derived from (**d**). **f** Magnetic phase diagrams for 206-nm-thick $FeGe^L$ and $FeGe^H$ films. The red area represents the phase space of the helimagnetic phase. The yellow area shows the part of the phase space that can be probed by REXS

subsequent measurements, we chose samples from two different growth series, differing in the substrate temperature during preparation, and characterised by a low T_c of ∼280 K (labelled $FeGe^L$) and a high T_c of ∼310 K ($FeGe^H$). Note that these two transition temperatures are dependent only on the growth conditions, and are largely independent of the film thickness in the range 40–200 nm.

Figure 2.10a shows the typical out-of-plane x-ray diffraction results for the as-grown FeGe thin films. Only the B20 FeGe (0,0,2) peak can be identified, which implies that the film is well-aligned with the substrate. No additional Fe-Ge phases are found. The full width at half maximum of the rocking curve about the FeGe (0,0,2) peak is less than 0.2°, as shown in the inset of Fig. 2.10a, indicating high crystalline quality. As shown in Fig. 2.10c, the magnetisation–temperature profile for a 200-nm-thick $FeGe^H$ sample resembles that found for FeGe bulk crystals, with a kink-like feature observed near the transition temperature in the zero-field-cooled curve implying a critical temperature $T_c \approx 310$ K. The M-H loop measured at 300 K, shown in Fig. 2.10b, is indicative of an anisotropic magnetically ordered state (where the anisotropy arises due to the weakly locked cubic anisotropy of bulk crystals being altered in the thin film limit, giving rise to an easy-plane uniaxial anisotropy). Figures 2.10d,e show M-H curves and their derivatives at different temperatures, which allow the magnetic phase boundary to be determined (Fig. 2.10e). The saturation magnetisation at 20 K is 0.77 μ_B/Fe, somewhat lower than the value reported for FeGe films by Porter et al. at $T = 5$ K [43] and less than the bulk value of 1.0 μ_B/Fe [45]. The critical fields and magnetic phase boundaries are similar to the ones previously reported for other B20 FeGe films on Si(111) [34, 42, 46]. The combined phase diagrams for $\sim$ 200-nm-thick FeGe^L and FeGe^H films are shown in Fig. 2.10f. Inside the boundaries, the films are helimagnetically ordered with the spin helix propagating along the film normal (as shown below).

The magnetic ground state structure of B20 compounds is a proper-screw-type spin spiral. It can be described by a one-dimensional harmonic model in which the spin rotates within a common plane that is perpendicular to the propagation direction, forming a periodic lattice structure that gives rise to magnetic Bragg reflections. However, the propagation direction of the wavevector $\mathbf{q}_h$ in thin film samples is found to be different from material to material [40, 47–50]. For example, for MnSi thin films, PNR and theoretical studies suggest that the spin helix propagates along the film normal, due to the enhanced uniaxial anisotropy [40, 49], while LTEM studies suggest that the helix is locked in-plane [47]. We performed REXS measurements on 200 nm FeGe^L and FeGe^H films in order to determine the helix propagation direction and the periodicity of the helical lattice.

Figure 2.11a shows the measurement geometry, where the incident light, with wavevector $\mathbf{k}_i$, is tuned to the Fe L_3 edge at 705 eV with σ-polarisation. The diffracted beam, with wavevector $\mathbf{k}_s$ (after scattering through wavevector $\mathbf{q}$), is captured either by a photodiode point detector or a CCD area detector as before. If the modulated spin structure is along the [001] direction (as shown), it will give rise to a magnetic diffraction peak along the l direction at $\mathbf{q}_h$ in reciprocal space. As $\mathbf{q}_h$ is small, the grazing incidence geometry is employed, with α_i scanning from 0° to 3.5°. However, various optical effects will occur in this geometry, such as specular reflectivity that satisfies $\alpha_i = \alpha_f$; the Yoneda peak that satisfies $\alpha_f = \theta_C$, where θ_C is the critical angle of the thin film; as well as the interference reflection from the thickness of the film that satisfies $2t \sin\alpha_i = n\lambda$, where t is the film thickness, and n is a positive integer. Therefore, special care needs to be taken in order to separate the magnetic peak from other non-magnetic contributions. This involves selecting a film thickness that is not a multiple of the helix pitch and also a systematic variation of parameters such as photon energy, sample temperature and applied field.

Figure 2.11b, c shows the direct CCD image for a FeGe^L sample at different temperatures, and Fig. 2.11d for a FeGe^H sample at 300 K. For the FeGe^L sample at 300 K no magnetic order is observable, therefore only the specular reflection and the Yoneda peak can be seen on the area detector. When cooled just below T_c, the magnetic diffraction peak appears which corresponds to the helical wavevector $\mathbf{q}_h$. Note that for a perfect diffraction condition, the magnetic peak $(0, 0, q_h)$ always overlaps with the specular reflectivity, and is therefore indistinguishable to the detector. However, in Fig. 2.11b–c, we intentionally offset the diffraction condition for the magnetic peak, leading to the separation between $(0, 0, q_h)$ and the specular. This peak corresponds to $l = q_h = 0.0063$ (r.l.u.). Reciprocal space scans along h at different temperatures are shown in Fig. 2.11f. Note that the non-zero background is

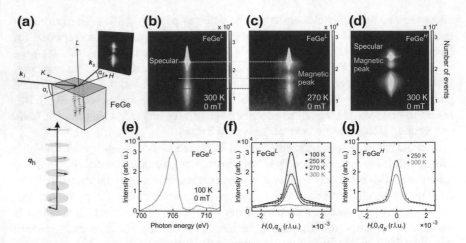

Fig. 2.11 Resonant elastic x-ray scattering on FeGe films. **a** Illustration of the scattering geometry. The incident (scattered) x-ray wavevectors are labelled as $\mathbf{k}_i$ ($\mathbf{k}_s$), and the corresponding incident (outgoing) angle α_i (α_f). The photon energy is tuned to 705 eV near the Fe L_3 edge. If $\mathbf{q}_h$ only propagates along the FeGe film normal, a magnetic peak at $(0, 0, q_h)$ is expected. However, this requires small incident angles. **b, c** CCD camera images showing the magnetic contrast at 300 and 270 K, at zero applied magnetic field for a FeGe^L film. α_i is 2.1° for both images. The reflectivity specular, Yoneda, and magnetics peaks are labelled. Note that the magnetic peak does not fulfil the perfect diffraction condition at this incident angle, therefore it separates from the specular peak. Further data processing reveals that this peak corresponds to $q_h = l = 0.0063$ (r.l.u.). **(d)** CCD camera image showing the magnetic $(0, 0, q_h)$ peak for the FeGe^H sample at 300 K ($\alpha_i = 1.5°$). Note that the magnetic peak is at a different position compared to (**c**) as the two α_i's are different. **e** Photon energy-dependence of the magnetic peak at $(0, 0, q_h)$ showing the magnetic origin of this peak. **f, g** Resonant h-scan about the magnetic satellite at different temperatures for both FeGe^L and FeGe^H samples

due to either diffuse scattering intensity from the specular and the Yoneda peaks, or the reflectivity diffusion from the thickness of the film. It has to be noted that only the modulation wavevector along l is observed, suggesting that the spins uniformly propagate along the film normal direction. This is in agreement with the conclusion reported in Refs. [40, 48, 49], in which the uniaxial anisotropy locks the propagation vector perpendicular to the film, regardless of its exact crystalline orientation. Secondly, we note that the corresponding real-space helical pitch is ∼74.6 nm, in agreement with the bulk behaviour [46]. Finally, the energy scan on $\mathbf{q}_h$, shown in Fig. 2.11e, confirms that the diffraction peak has an ordered magnetic origin.

Most significantly, the same magnetic peak that corresponds to the FeGe helical ground state is observed in the FeGe^H sample at room temperature, as shown in the CCD image in Fig. 2.11d and the h-scan in Fig. 2.11g, suggesting the successful realisation of room temperature helimagnetism in B20 compound thin films.

References

1. J. Als-Nielsen, D. McMorrow, *Elements of Modern X-ray Physics* (Wiley, 2010)
2. G. van der Laan, C. R. Phys. **9**, 570 (2008)
3. P.M. Platzman, N. Tzoar, Phys. Rev. B **2**, 3556 (1970)
4. F. de Bergevin, M. Brunel, Acta Cryst. A **37**, 314 (1981)
5. M. Blume, D. Gibbs, Phys. Rev. B **37**, 1779 (1988)
6. S.W. Lovesey, S.P. Collins, *X-ray Scattering and Absorption by Magnetic Materials* (Oxford University Press, 2002)
7. B.T. Thole, G. van der Laan, G.A. Sawatzky, Phys. Rev. Lett. **55**, 2086 (1985)
8. G. van der Laan, B.T. Thole, G.A. Sawatzky, J.B. Goedkoop, J.C. Fuggle, J.-M. Esteva, R. Karnatak, J.P. Remeika, H.A. Dabkowska, Phys. Rev. B **34**, 6529 (1986)
9. M. Blume, J. Appl. Phys. **57**, 3615 (1985)
10. D. Gibbs, D.R. Harshman, E.D. Isaacs, D.B. McWhan, D. Mills, C. Vettier, Phys. Rev. Lett. **61** (1988)
11. J.P. Hannon, G.T. Trammell, M. Blume, D. Gibbs, Phys. Rev. Lett. **61**, 1245 (1988)
12. G. van der Laan, B.T. Thole, Phys. Rev. B **43**, 13401 (1991)
13. B.T. Thole, G. van der Laan, J.C. Fuggle, G.A. Sawatzky, R.C. Karnatak, J.-M. Esteva, Phys. Rev. B **32**, 5107 (1985)
14. N. Nagaosa, Y. Tokura, Nat. Nanotech. **8**, 899 (2013)
15. J.-W.G. Bos, C.V. Colin, T.T.M. Palstra, Phys. Rev. B **78**, 094416 (2008)
16. G. van der Laan, A.I. Figueroa, Coord. Chem. Rev. **277–278**, 95 (2014)
17. S. Grenier, Y. Joly, J. Phys. Conf. Ser. **519**, 012001 (2014)
18. J. Fink, E. Schierle, E. Weschke, J. Geck, Rep. Prog. Phys. **76**, 056502 (2013)
19. G. van der Laan, Phys. Rev. B **86**, 035138 (2012)
20. J.P. Hill, D.F. McMorrow, Acta Cryst. Sect. A **52**, 236 (1996)
21. D.H. Templeton, L.K. Templeton, Phys. Rev. B **49**, 14850 (1994)
22. T. Adams, A. Chacon, M. Wagner, A. Bauer, G. Brandl, B. Pedersen, H. Berger, P. Lemmens, C. Pfleiderer, Phys. Rev. Lett. **108**, 237204 (2012)
23. J.S. White, K. Pršsa, P. Huang, A.A. Omrani, I. Živkovi ć, M. Bartkowiak, H. Berger, A. Magrez, J.L. Gavilano, G. Nagy, J. Zang, H.M. Rnnow, Phys. Rev. Lett. **113**, 107203 (2014)
24. T.A.W. Beale, T.P.A. Hase, T. Iida, K. Endo, P. Steadman, A.R. Marshall, S.S. Dhesi, G. van der Laan, P.D. Hatton, Rev. Sci. Instrum. **81**, 073904 (2010)
25. A.J. Schwartz, M. Kumar, B.L. Adams, D.P. Field (eds.), *Electron Backscatter Diffraction in Materials Science* (Springer, 2009)
26. W. Munzer, A. Neubauer, T. Adams, S. Muhlbauer, C. Franz, F. Jonietz, R. Georgii, P. Boni, B. Pedersen, M. Schmidt, A. Rosch, C. Pfleiderer, Phys. Rev. B **81**, 041203(R) (2010)
27. P. Milde, D. Kohler, J. Seidel, L.M. Eng, A. Bauer, A. Chacon, J. Kindervater, S. Muhlbauer, C. Pfleiderer, S. Buhrandt, C. Schutte, A. Rosch, Science **340**, 1076 (2013)
28. H. Oike, A. Kikkawa, N. Kanazawa, Y. Taguchi, M. Kawasaki, Y. Tokura, F. Kagawa, Nat. Phys. **12**, 62 (2016)
29. T. Adams, S. Muhlbauer, A. Neubauer, W. Munzer, F. Jonietz, R. Georgii, B. Pedersen, P. Boni, A. Rosch, C. Pfleiderer, J. Phys. Conf. Ser. **200**, 032001 (2010)
30. S.L. Zhang, A. Bauer, D.M. Burn, P. Milde, E. Neuber, L.M. Eng, H. Berger, C. Pfleiderer, G. van der Laan, T. Hesjedal, Nano Lett. **16**, 3285 (2016)
31. B.T. Thole, G. van der Laan, Phys. Rev. B **38**, 3158 (1988)
32. S. Seki, X.Z. Yu, S. Ishiwata, Y. Tokura, Science **336**, 198 (2012)
33. S. Muhlbauer, B. Binz, F. Jonietz, C. Pfleiderer, A. Rosch, A. Neubauer, R. Georgii, P. Boni, Science **323**, 915 (2009)
34. Y. Yamasaki, D. Morikawa, T. Honda, H. Nakao, Y. Murakami, N. Kanazawa, M. Kawasaki, T. Arima, Y. Tokura, Phys. Rev. B **92**, 220421 (2015)
35. I.K. Robinson, Phys. Rev. B **33**, 3830 (1986)
36. J.R. Levine, J.B. Cohen, Y.W. Chung, P. Georgopoulos, J. Appl. Cryst. **22**, 528 (1989)

37. G. Renaud, R. Lazzari, F. Leroy, J. Surfrep. **64**, 255 (2009)
38. A. Tonomura, X.Z. Yu, K. Yanagisawa, T. Matsuda, Y. Onose, N. Kanazawa, H.S. Park, Y. Tokura, Nano Lett. **12**, 1673 (2012)
39. E. Magnano, E. Carleschi, A. Nicolaou, T. Pardini, M. Zangrando, F. Parmigiani, Surf. Sci. **600**, 3932 (2006)
40. E.A. Karhu, S. Kahwaji, M.D. Robertson, H. Fritzsche, B.J. Kirby, C.F. Majkrzak, T.L. Monchesky, Phys. Rev. B **84**, 060404(R) (2011)
41. S.L. Zhang, R. Chalasani, A.A. Baker, N.-J. Steinke, A.I. Figueroa, A. Kohn, G. van der Laan, T. Hesjedal, AIP Adv. **6**, 015217 (2016)
42. S.X. Huang, C.L. Chien, Phys. Rev. Lett. **108**, 267201 (2012)
43. N.A. Porter, J.C. Gartside, C.H. Marrows, Phys. Rev. B **90**, 024403 (2014)
44. N.A. Porter, C.S. Spencer, R.C. Temple, C.J. Kinane, T.R. Charlton, S. Langridge, C.H. Marrows, Phys. Rev. B **92**, 144402 (2015)
45. H. Yamada, K. Terao, H. Ohta, E. Kulatov, Phys. B **329–333**, 1131 (2003)
46. X.Z. Yu, N. Kanazawa, Y. Onose, K. Kimoto, W.Z. Zhang, S. Ishiwata, Y. Matsui, Y. Tokura, Nat. Mater. **10**, 106 (2011)
47. Y. Li, N. Kanazawa, X.Z. Yu, A. Tsukazaki, M. Kawasaki, M. Ichikawa, X.F. Jin, F. Kagawa, Y. Tokura, Phys. Rev. Lett. **110**, 117202 (2013)
48. M.N. Wilson, E.A. Karhu, D.P. Lake, A.S. Quigley, A.N. Bogdanov, U.K. Roler, T.L. Monchesky, Phys. Rev. B **88**, 214420 (2013)
49. M.N. Wilson, A.B. Butenko, A.N. Bogdanov, T.L. Monchesky, Phys. Rev. B **89**, 094411 (2014)
50. N. Kanazawa, M. Kubota, A. Tsukazaki, Y. Kozuka, K.S. Takahashi, M. Kawasaki, M. Ichikawa, F. Kagawa, Y. Tokura, Phys. Rev. B **91**, 041122(R) (2015)

Chapter 3
Measurement of the Skyrmion Lattice Domains

So far we have dealt two different resonant x-ray diffraction geometries, and each of them has its unique features that make them a stand-alone technique. The simplest variation of such a diffraction technique is to utilise the diffraction pattern obtained from a focused beam as a local probe, and perform a scanning probe type measurement to map out the two-dimensional space. The analysis of the images demands the assistance of pattern-recognition algorithms. One example is the electron back-scatter diffraction (EBSD) method that does imaging on crystalline domains [1]. Given that the electron beam can be focused into a very small spot, which is smaller than a single domain of a polycrystalline specimen, the electrons 'see' at each pixel position a 'single crystal', which gives rise to a characteristic diffraction pattern. By carrying out a raster scan, the diffraction pattern changes from domain to domain. Eventually one obtains a two-dimensional map, where each pixel corresponds to a diffraction pattern which is to be further analysed to determine its crystalline orientation. Then the real-space domain information can be reconstructed.

What we are going to do in this chapter is to apply this concept to REXS. The most important aspect that evaluates whether this is feasible is the ratio between the averaged single domain size and the local probe resolution, i.e., the beam size. For our instrumental setup, the smallest beam that can be obtained is a 20 μm-diameter circular area, limited by a pinhole, which also makes the beam coherent. It is small enough to carry out a two-dimensional raster scan on the skyrmion lattice phase, to look for differently oriented domains. Next, the question of pattern-recognition has to be addressed, i.e., how to define a single domain based on the diffraction pattern? Fortunately this is rather simple in our case, as a single skyrmion lattice domain only contains six well-defined magnetic peaks. If the hexagonally-ordered lattice rotates in plane, the associated diffraction pattern will rotate accordingly. This fact largely reduces the computing power necessary to process the image data.

© Springer Nature Switzerland AG 2018
S. Zhang, *Chiral and Topological Nature of Magnetic Skyrmions*,
Springer Theses, https://doi.org/10.1007/978-3-319-98252-6_3

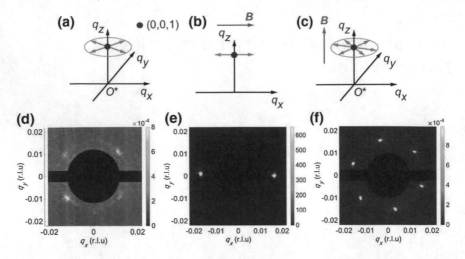

Fig. 3.1 Typical REXS data on different magnetic orders in Cu_2OSeO_3 single crystal. Magnetic satellites in reciprocal space for **a** the helical (15 K, 0 mT), **b** conical (50 K, 32 mT, $\gamma = 90°$), and **c** skyrmion lattice state (57 K, 32 mT, $\gamma = 0°$), respectively. **d–f** Reciprocal space maps of the helical, conical, and skyrmion orders, respectively, using π-polarized incidence light with a photon energy of 931.25 eV. The plot is the hk plane at $l = 1$ in reciprocal space

3.1 Diffraction from the Multidomain Skyrmion Lattice State

The best geometry to demonstrate this imaging technique is reflection REXS on Cu_2OSeO_3, as the $\sim 48.2°$ incidence angle has a very well-defined beam shape on the sample. On the other hand, if grazing-incidence is used, the beam will be largely elongated on the sample surface, which is not suitable for a raster scan.

Therefore, let us start with typical REXS data of this sample, as shown in Fig. 3.1. Note that this time π-polarisation is used, making the diffraction brighter. This is due to the polarisation-dependence of REXS, which will be discussed in Chap. 4. It is worth emphasising that the beam size used here is 0.5×0.5 mm². For such probing area, a well-defined six-fold-symmetric skyrmion lattice peak is observed at $\gamma = 0°$, suggesting a long-range-ordered, single-domain skyrmion phase.

Next, we focus on the skyrmion lattice where Fig. 3.2 presents the integrated CCD images for different tilt of the magnetic field γ. Starting well above T_c at 65 K, a field of 32 mT is applied under a fixed γ. After cooling the sample to 56.5 K and stabilising the temperature for 15 min, a RSM is carried out. After the scan, the sample is heated to 65 K. Up to a tilt of $\gamma = -2°$, see Fig. 3.2a, the six-fold-symmetric diffraction pattern remains unchanged with one of the Friedel pairs essentially aligned along h. For a further increased γ, see Fig. 3.2b, the sixfold pattern reorients. Note that for $\gamma = -5°$ the skyrmion plane is tilted by $\sim 5°$ with respect to the crystalline [001].

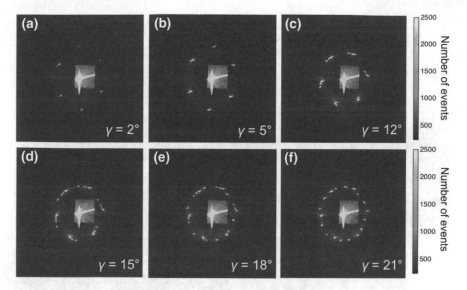

Fig. 3.2 Sum of CCD images in the skyrmion lattice phase at different field angles γ (**a–f**). The sample was field-cooled in 32 mT at the indicated γ from 65 K down to 56.5 K before a RSM was recorded. As a function of increasing γ the single sixfold scattering pattern prevalently splits into several sixfold symmetric subsets. Reprinted with permission from Ref. [2]. Copyright 2016 by American Chemical Society

As a result, no easy axes, as defined by the cubic anisotropy, are available for the propagation vectors perpendicular to the field direction. Therefore, the propagation vectors only roughly follow the projected easy axes directions.

As shown in Fig. 3.2c–f, for larger values of γ, the sixfold intensity distribution splits up into several sixfold subsets that become increasingly pronounced with increasing γ. For $\gamma = -21°$, the diffraction pattern finally resembles a necklace-like diffraction pattern. The intensity distributions can be reproduced by repeating the same temperature-field history. Positive as well as negative values of γ lead to the same results. As a function of time, the intensity distributions stay unchanged for at least one hour. Moreover, once a split pattern has formed, the rotation of γ back to $0°$ at constant temperature and magnetic field does not alter the pattern (on a timescale of at least 15 mins).

The REXS patterns for different states of the skyrmion lattice, e.g., single versus multidomain, can be expected to be different if the domains are smaller than the beam size of the x-rays. Figure 3.3a shows the spin configuration for an individual skyrmion vortex as the motif of the hexagonal lattice in the single-domain skyrmion state (shown in Fig. 3.3b). The calculated REXS pattern is shown in Fig. 3.3c. Note the diffraction pattern does not change with different x-ray polarisations, only the intensity distribution among the diffraction peaks will change. The calculated pattern is in excellent agreement with our experimental results (see, e.g., Fig. 3.2f).

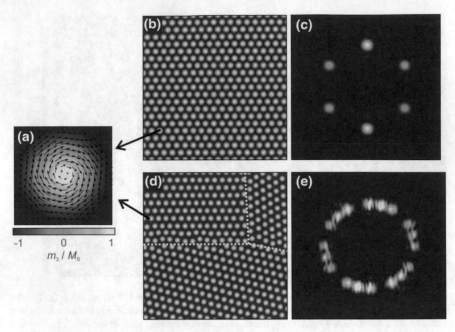

Fig. 3.3 Numerical calculations of the REXS signal for different skyrmion lattice states. **a** Magnetisation configuration of an individual skyrmion that is used as the motif to construct the lattice for the calculations. The colour scale represents the z-component of the magnetisation unit vector. **b** Single domain skyrmion lattice used for the simulation. **c** Reciprocal space REXS pattern for σ-polarised x-rays at the Cu L_3 edge. **d** Real-space multidomain skyrmion state. **e** Simulated REXS pattern using the same incident x-rays as in (**c**). Reprinted with permission from Ref. [2]. Copyright 2016 by American Chemical Society

Next, three domains are sampled in the same area, i.e., a multidomain state. The domain boundaries are indicated by dashed yellow lines (Fig. 3.3d). Within each domain, a well-defined hexagonal skyrmion lattice is found that is rotated with respect to the neighbouring domains. The REXS pattern is simulated in Fig. 3.3e, which recovers the necklace-like pattern as experimentally observed in Fig. 3.2. The other scenario can also satisfy the experimental observation, however keeping the lateral single domain state. In this scenario, the three-dimensional skyrmion tubes that are along the field direction break up into several layers, that are differently oriented in-plane. However, this scenario will generate great magnetisation discontinuities and artificial topological defects that cost much more energy, leading to an extremely unstable state. Therefore, this structure interpretation can be ruled out, purely based on the diffraction experiments. It is significant that the in-plane orientation of the domains shows no clear preferred locking direction as imposed by the cubic anisotropy, and the fragmentation is far more pronounced than in other chiral magnets [3, 4]. Moreover, the formation of these domains can be reproducibly induced by a tilted magnetic field as the tuning parameter.

Here we will discuss the possible origins of the formation of a multidomain skyrmion lattice state. As we have shown, REXS only probes the magnetic order on the surface and surface-near areas of the material. Note that the magnetic state on the surface may be quite different from the bulk due to the influence of surface anisotropy. The total energy density of such system can be again referred to Eq. (2.26), which we repeat here for convenience:

$$w(m) = A(\nabla m)^2 + Dm \cdot (\nabla \times m) - B \cdot m + w_A , \qquad (3.1)$$

Here, the second term in Eq. (3.1) takes different form on the surface out of a three-dimensional bulk system. The extra boundary conditions may induce a completely different skyrmion structure. The detailed surface energy term and its induced magnetic structures are theoretically discussed in [5–10]. This may be the most important ingredient for the modification of the surface skyrmion lattice state observed in our surface-sensitive REXS experiments.

On the other hand, for $B = 0$, the helical ground state minimises the total energy, and using the inverse Fourier transform, the magnetisation is rewritten as $m(r) = M(q_h)e^{iq_h \cdot r}$ + c.c., where c.c. denotes the complex conjugate. Note $M(q) = 0$ for $q \neq q_h$, therefore only the Fourier components at the diffraction peaks can be considered when reconstructing the real-space magnetisation profiles.

Using the equivalence of the axial-symmetric skyrmion lattice solution and the triple-q spin density wave solution for Eq. (2.26) (with the corrections of the thermal fluctuation), at finite fields, the skyrmion lattice can also take the form of

$$m(r) = \frac{1}{3} \sum_{i=1}^{3} \left[M(q_i)e^{iq_i \cdot r} + \text{c.c.} \right] + m_{\text{net}} , \qquad (3.2)$$

where q_1, q_2, and q_3 coherently propagate in the plane perpendicular to B, taking the form $q_i = q[\hat{q}_x \cos(\Psi_i) + \hat{q}_y \sin(\Psi_i)]$. $\hat{q}_x$ and $\hat{q}_y$ are the orthogonal unit vectors of the two-dimensional reciprocal space; Ψ_1, Ψ_2, and Ψ_3 describe the azimuthal angles, which are 120° apart from each other, and m_{net} provides the ferromagnetic background, in which only m_3 is shown with a constant value (One can actually take $m_{\text{net}} = 0$ if the measured net magnetisation is zero). The advantage of using this triple-q solution is that the skyrmion wavevector locking mechanism is simply represented by Ψ_i.

If w_A is switched off, Eq. (3.1) perseveres the rotational symmetry, meaning that Ψ_i can be locked into an arbitrary direction. On the other hand, if the cubic anisotropy shows up due to the cubic crystalline environment, the azimuthal angles Ψ_i will depend on the sixth-order cubic anisotropy term [11, 12] for Cu_2OSeO_3, and read

$$w_A^{\text{cubic}(6)} = am_1^2 m_2^2 m_3^2 + b(m_1^2 m_2^4 + m_2^2 m_3^4 + m_3^2 m_1^4)$$
$$+ c(m_2^2 m_1^4 + m_3^2 m_2^4 + m_1^2 m_3^4) + d(m_1^6 + m_2^6 + m_3^6) , \qquad (3.3)$$

where a, b, c, and d are material-dependent amplitude constants. By substituting this into Eq. (3.1) the minimum energy is obtained when one of the three Ψ_i is pinned along a $\langle 100 \rangle$ direction, while being a solution of Eq. (3.2) for the skyrmion lattice.

We consider a surface anisotropy different from the bulk anisotropy, which extends up to a few unit cells in depth and which may be induced by symmetry breaking. We assume that across the depth of a few unit cells, the expression simplifies to [13]

$$w_A^{\text{surface}} = K_{\text{u}} m_3^2 . \tag{3.4}$$

The uniaxial anisotropy constant K_{u} is positive, describing an easy-plane anisotropy, similar to the case of MnSi thin films [8, 14]. By inserting Eq. (3.4) into Eq. (3.1), one obtains that the energy term possesses $SO(2)$ symmetry, the group of rotations about the B direction. Therefore, the Ψ_i are not pinned, suggesting that all q_i propagation directions in the plane perpendicular to B are degenerate.

In our REXS experiment, soft x-rays probe $\sim$38 unit cells, i.e., much deeper than the extend of the surface anisotropy. At this depth, the system is governed by the competition between w_A^{cubic} and w_A^{surface}. For $B \parallel [001]$, the cubic anisotropy dominates over the surface anisotropy and locks Ψ_i, as observed in Fig. 3.1c. When tilting the field B by an angle γ an in-plane component of the field arises, inducing a depinning effect of the skyrmion lattice from the original $\langle 100 \rangle$ direction. With increasing γ, the $\langle 001 \rangle$ locking protocol does not apply due to the geometry change. On the other hand, the depinning of the skyrmion lattice allows the surface anisotropy to become dominant, therefore the propagation direction becomes arbitrary. Eventually a multidomain skyrmion state is formed, as observed in Fig. 3.2.

Furthermore, another important ingredient that may be responsible for the multidomain skyrmion state is the magnetoelectric coupling that separates Cu_2OSeO_3 from other cubic chiral magnets. Calculations suggest that a magnetic field along $\langle 100 \rangle$ results in low overall values of the local electric polarisation [15], consistent with a single-domain skyrmion lattice for fields along [001]. On the other hand, magnetic fields along other directions lead to large in-plane or out-of-plane electric dipole moments. The complex interplay of the magnetic and electric dipole moments, in combination with the delicate cubic anisotropy [16], may finally induce the formation of multiple domains. This assumption is also corroborated by the reorientation of the skyrmion lattice in external electrical fields [17]. Moreover, note that in our measurements the multidomain pattern is observed for exposure times of only 2 ms, and it does not evolve over time. Therefore, we also exclude a dynamic origin that was introduced as the explanation for a double-split pattern observed in Lorentz transmission electron microscopy on thinned plates of Cu_2OSeO_3 for long, averaging exposure times of 100 ms [4].

3.2 Imaging of the Multidomain Skyrmion Lattice State

From the previous diffraction experiments, there is one fact that worth noticing: the necklace-like pattern contains 'beads', and the number of the 'beads' is countable. This implies that the multidomain skyrmion lattice state has a very limited number of the domains within the probing area. Therefore, the domain size is a large fraction of the beam size. This fact directly hints that if the beam size can be reduced by more than an order of magnitude (which is possible), a diffraction imaging method can be used to map out the real-space domain distributions.

The sample shape and geometry we used here is shown Fig. 3.4a, in which the raster scan area of 1×1 mm^2 is marked by the blue square. We first recover the multidomain skyrmion lattice state using the field-cooling protocol introduced before. As shown in Fig. 3.4b, by 32 mT, $\gamma = 15°$ field cool down from 65 to 57 K, the necklace-like diffraction pattern is observed. Next, for the same experimental conditions, by inserting the 20 µm-diameter pinhole in front of the sample, the beam size shrinks accordingly, and the x-rays become coherent. The distance between the pinhole and the sample is 26.5 mm, which gives rise to Fresnel diffraction [18] if the pinhole diameter is less than 5 µm. In such a near-field limit, the diffracted beam from the pinhole cannot be approximated by plane waves, therefore it is not compatible with our theoretical descriptions in Chap. 2. In order to achieve the Fraunhofer diffraction (i.e., the far-field diffraction limit) [18], a pinhole size of larger than 20 µm has to be used. This determines the resolution of our scanning diffraction imaging. The direct consequence of the beam-shrinking is that the necklace-like pattern suddenly turns into a single-domain, six-fold-symmetric pattern, as shown in Fig. 3.4c. This confirms that within the area of 100π µm^2, the skyrmion lattice is single domain. It again provides hard evidence that the interpretation of the lateral skyrmion domain based on the diffraction result is valid.

Then the raster scan using the 20 µm beam is performed on the same 1×1 mm^2 surface lateral area for different skyrmion lattice domain states. The different states

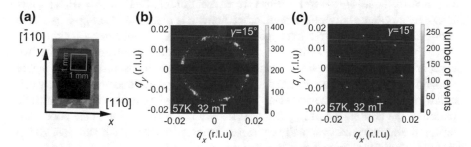

Fig. 3.4 Characterisation of the skyrmion lattice domains. **a** Photo of the sample and its orientation. The blue square marks the real-space region mapped using diffraction imaging. **b** Reciprocal space map of the skyrmion plane that is perpendicular to the external field, reached by field cooling down to 57 K in a field of 32 mT and $\gamma = 15°$. The sampling area is 300×300 µm^2. **c** Skyrmion plane reciprocal space map using a reduced beam size of 20 µm in diameter, but otherwise identical conditions as in (**b**)

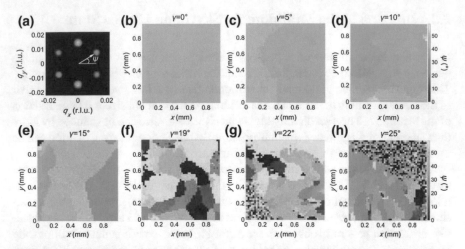

Fig. 3.5 REXS scan imaging results. **a** Legend of the domain image, in which $\Psi = 0°$ is defined as one pair of the magnetic peaks oriented along q_x direction. Note that Ψ varies from $0°$ to $60°$. **b–h** Systematic raster scan images showing the domain maps that correspond to the sample area in the blue box in Fig. 3.4a

are generated by field cooling down from 65 to 57 K in a field of 32 mT with a specific γ angle. When stabilising for 15 mins, the areal scan is carried out. Therefore, each image contains 50×50 pixels. At most of the pixels, a six-fold-symmetric, single domain diffraction pattern can be observed. At certain pixels, at which the domain boundaries exist, a multitude of the six-fold-symmetric pattern can be found. We therefore process the camera image at each pixel, and extract the lattice orientation angle Ψ. Then the domain image can be reconstructed. Figure 3.5a shows the legend of Ψ, and $\Psi = 0°$ is defined as one pair of the skyrmion magnetic peaks align along q_x.

The evolution of the domain pattern at different field titling angle is shown in Fig. 3.5b–h. There are several features worth mentioning. First, at $\gamma = 0°$, a perfect long-range-ordered single skyrmion lattice domain is observed (see Fig. 3.5a). This is consistent with the diffraction experiments, as well as the anisotropy-locking theory. When the field slightly tilts, the domains start to evolve, manifesting itself as two compartments, with the major orientation approximately the same as for the single domain case. When further increasing the field, the domains become randomly oriented, and the average size decreases with increasing γ. As γ goes to $22°$, a mosaic pattern at the bottom-left of Fig. 3.5g can be observed. This is due to the poor resolution: the domain size eventually decreases down to less than the beam size, giving rise to a multiple-domain diffraction pattern at each pixel.

Second, the shape of each domain is rather irregular, and the distribution of them is random. This suggests that the domain formation is spontaneous, and not due to defect-pinning. The domain pattern at each γ is stable, confirmed by seeing identical images from the raster scans on the same area when repeated multiple times. However, for the same γ, if the temperature increases above T_c, and cools down again, the image will show a completely different domain pattern.

Third, the orientations of the neighbouring domains intend to have a 'gradual' transition across the boundary, i.e., the average Ψ difference between the two neighbouring domains is usually less than $10°$. At smaller γ angles, the cross-over between domains is more gradual, some of which even have a continuous rotation from one to another, see for example Fig. 3.5d. However, at larger titling angles, a sharp rotation of the lattices across boundaries becomes common.

This 'polycrystalline' type of the skyrmion lattice, and its real-space appearance, have not been observed before. It shows the delicate balance of the system's energy hierarchy, in which multiple interactions compete. The isotropic exchange interaction, DMI and Zeeman energy decide the major magnetic order, while anisotropy, demagnetisation, and possibly the ferroelectric effect serve as perturbations, which do the finer adjustment of the system's structure. Another interpretation of these phenomena can be correlated with topological effects. As the skyrmion vortices are topologically stable states, thermal fluctuations actually stabilise them, instead of destroying them. Therefore, once the state is formed, it intends to survive other energy perturbations. For example, if there is a structural defect on the surface (which there always are), it will manifest itself as a magnetic local defect. Even so, the skyrmions will adapt to such an environment by adjusting the vortex shape locally, and on the large scale, the lattice will completely ignore such local defects as a long-wavelength spin density wave. (Think about how a water wave 'ignores' small rocks during its propagation). In another instance, if an extra anisotropy plays a role in the system (which is also true for the surface), the skyrmions will also try to keep their topological configuration, while adapting to such an environment by breaking up into domains, instead of unwinding into other spin configurations.

3.3 Domain Boundary Structure

In the last section, I would like to discuss some possible structures at the domain boundary. This information cannot be measured by REXS, but microscopy has to be performed to map out the local spin configurations.

From this perspective, LTEM is the most suitable technique for studying the magnetic domain boundary structure between two skyrmion lattice domains. However, due to its small field of view, one has to be very patient to search for the domain areas. Figure 3.6 shows LTEM images that reveal this information. As shown in Fig. 3.6a–c, in order to break up into differently oriented lattices, the closed-pack ordering must have defects at the boundaries, i.e., there are five or seven skyrmions surrounding a reference skyrmion, instead of six [4]. In this case, a line defect can appear, forming the domain boundary. Depending on its configuration, the neighbouring domains can change their orientations to have an arbitrary angle, depending on the width of the defect-containing boundary. The thicker this boundary is, the large the relative rotation will be. If we relate this to our domain observations, it points towards the

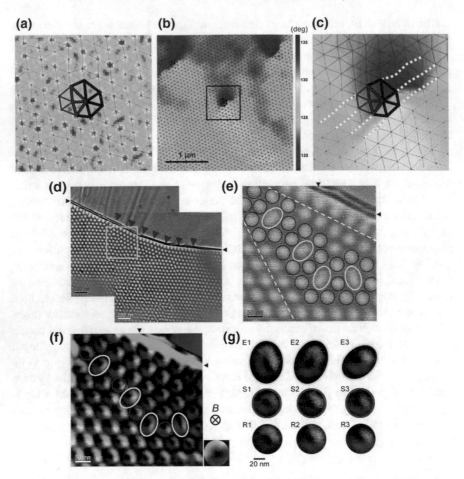

Fig. 3.6 Lorentz transmission electron microscopy images for the skyrmion lattice phase in **a–c** Cu_2OSeO_3. Reprinted with permission from Ref. [4]. Copyright 2016 by National Academy of Sciences. **d–g** LTEM images for thinned MnFeGe specimen prepared from bulk samples. Reprinted from Ref. [19]. Copyright 2016 The Authors and American Association for the Advancement of Science. The dislocations of the long-range-ordered skyrmion lattice are captured in these images, from which the skyrmion domain boundary structure is characterised

fact that under small γ, the domain boundary contains thin defect lines, leading to small relative rotations. With increasing γ, the number of defects increases across the domain boundary, giving rise to sharp rotational transitions.

The other aspect of the local defect structure is the shape of the skyrmions. As these topologically-protected entities behave like quasi-particles, they prefer to keep the closed-packing order, i.e., less ferromagnetic background is preferred. However, if a five- or seven-neighbour scenario is introduced, more 'empty' space will be left out for the ferromagnetic if the skyrmions' shape and size are kept the same. This apparently contradicts to the energy minimisation preference of the system.

As a result, the skyrmions in the defect region are distorted, in order to occupy the 'empty' space. This is also observed in LTEM, as shown in Fig. 3.6d–g. At the domain boundary of the five- or seven-neighbour regions, elliptical skyrmions can be identified [19]. The topological winding number for these distorted skyrmions is the same as for standard ones, implying the robustness of such quasi-particles.

To briefly summarise this chapter, we have demonstrated a technique derived from REXS, that can achieve skyrmion lattice domain imaging. This method provides important information on the domain distribution, shape, size, and formation. These findings may enable an effective solutions for skyrmion-based applications. First, the skyrmions are robust, meaning that they are good information carriers. However, one can never use a millimetre-sized, single domain skyrmion lattice [20] as a bit of information. Instead, the long range order has to be broken up into smaller parts, while the individual parts can be served as bits. This was shown to be possible by introducing extra perturbations. Second, such individual domains are required to be as small as possible. The extreme case is that each domain contains one skyrmion vortex. This can be achieved by using the tuning parameter of γ—at least we have demonstrated that the minimum domain size can be smaller than 20 μm.

References

1. A.J. Schwartz, M. Kumar, B.L. Adams, D.P. Field (eds.), *Electron Backscatter Diffraction in Materials Science* (Springer, 2009)
2. S.L. Zhang, A. Bauer, D.M. Burn, P. Milde, E. Neuber, L.M. Eng, H. Berger, C. Pfleiderer, G. van der Laan, T. Hesjedal, Nano Lett. **16**, 3285 (2016)
3. T. Adams, S. Mühlbauer, A. Neubauer, W. Münzer, F. Jonietz, R. Georgii, B. Pedersen, P. Böni, A. Rosch, C. Pfleiderer, J. Phys. Conf. Ser. **200**, 032001 (2010)
4. J. Rajeswaria, H. Pinga, G.F. Mancini, Y. Murooka, T. Latychevskaia, D. McGrouther, M. Cantoni, E. Baldini, J.S. White, A. Magrez, T. Giamarchi, H.M. Rønnow, F. Carbone, Proc. Natl. Acad. Sci. U.S.A. **112**, 14212 (2015)
5. S. Rohart, A. Thiaville, Phys. Rev. B **88**, 184422 (2013)
6. F.N. Rybakov, A.B. Borisov, A.N. Bogdanov, Phys. Rev. B **87**, 094424 (2013)
7. M.N. Wilson, E.A. Karhu, D.P. Lake, A.S. Quigley, A.N. Bogdanov, U.K. Rößler, T.L. Monchesky, Phys. Rev. B **88**, 214420 (2013)
8. S.A. Meynell, M.N. Wilson, H. Fritzsche, A.N. Bogdanov, T.L. Monchesky, Phys. Rev. B **90**, 014406 (2014)
9. J. Müller, A. Rosch, M. Garst, New J. Phys. **18**, 065006 (2016)
10. A.O. Leonov, Y. Togawa, T.L. Monchesky, A.N. Bogdanov, J. Kishine, Y. Kousaka, M. Miyagawa, T. Koyama, J. Akimitsu, T. Koyama, K. Harada, S. Mori, D. McGrouther, R. Lamb, M. Krajnak, S. McVitie, R.L. Stamps, K. Inoue, Phys. Rev. Lett. **117**, 8 (2016)
11. S. Mühlbauer, B. Binz, F. Jonietz, C. Pfleiderer, A. Rosch, A. Neubauer, R. Georgii, P. Böni, Science **323**, 915 (2009)
12. J.S. White, K. Prša, P. Huang, A.A. Omrani, I. Živković, M. Bartkowiak, H. Berger, A. Magrez, J.L. Gavilano, G. Nagy, J. Zang, H.M. Rønnow, Phys. Rev. Lett. **113**, 107203 (2014)
13. A. Hubert, R. Schäfer, *Magnetic Domains—The Analysis of Magnetic Microstructures* (Springer, 2008)
14. E.A. Karhu, U.K. Rößlerler, A.N. Bogdanov, S. Kahwaji, B.J. Kirby, H. Fritzsche, M.D. Robertson, C.F. Majkrzak, T.L. Monchesky, Phys. Rev. B **85**, 094429 (2012)
15. S. Seki, S. Ishiwata, Y. Tokura, Phys. Rev. B **86**, 060403(R) (2012)

16. S. Seki, J.-H. Kim, D.S. Inosov, R. Georgii, B. Keimer, S. Ishiwata, Y. Tokura, Phys. Rev. B **85**, 220406(R) (2012)
17. J.S. White, I. Levatić, A.A. Omrani, N. Egetenmeyer, K. Prša, I. Živković, J.L. Gavilano, J. Kohlbrecher, M. Bartkowiak, H. Berger, H.M. Rønnow, J. Phys. Cond. Matter **24**, 432201 (2012)
18. J. Als-Nielsen, D. McMorrow, *Elements of Modern X-ray Physics* (Wiley, 2010)
19. T. Matsumoto, Y.-G. So, Y. Kohno, H. Sawada, Y. Ikuhara, N. Shibata, Sci. Adv. **2**, e1501280 (2016)
20. T. Adams, S. Mühlbauer, C. Pfleiderer, F. Jonietz, A. Bauer, A. Neubauer, R. Georgii, P. Böni, U. Keiderling, K. Everschor, M. Garst, A. Rosch, Phys. Rev. Lett. **107**, 217206 (2011)

Chapter 4
Measurement of the Topological Winding Number

So far, we have focused on the incommensurate, modulated spin structures by study-ing the diffraction peak positions in reciprocal space. This gives us the symmetry of the magnetic order, from which the different phases can be distinguished. The absolute length of these magnetic peaks reveals the modulation periodicity, while the direction of these peaks reveal the information of the anisotropy. Based on these relationships, the properties that originate from long-range order are studied. How-ever, the diffraction intensity of these magnetic peaks are overlooked by far.

From now on, we are going to focus on the details of the diffraction structure fac-tors, especially their polarisation dependence. As can be expected, these reveal the information about the motifs that constitute the periodic structures. For the skyrmion lattice phase, the motif is of course the individual skyrmion vortex, which is of great interest. We will start by looking at the most 'insensitive' aspect of the skyrmions: their winding number. The reason why we call it insensitive is due to the topological properties: even if local surgery is performed on a skyrmion vortex, the winding num-ber is not changed, leaving the experimental data unchanged. From this perspective, such a technique should not be limited to skyrmion-carrying materials. Instead, it is a more general method that is suitable to study the topological properties of many magnetic materials.

The mathematical concept of topology deals with the fundamental properties of the order parameter space, and provides insights for the study of geometrical prop-erties and spatial relations unaffected by the continuous change of shape or size of figures [1]. It brings about significant advantages that allow for a fundamental understanding of the underlying physics of, e.g., superfluids [2], liquid crystals [3], superconductors [4], topological insulators [5], and magnetic materials [6]. In mag-netism, the topology of spin order manifests itself in the topological winding number [7]. It plays a pivotal role in the determination of a system's emergent properties leading, e.g., to potential spintronic applications in the recently discovered mag-netic skyrmion materials [8]. However, the direct experimental determination of the topological winding number of a magnetically ordered system remains challenging.

© Springer Nature Switzerland AG 2018
S. Zhang, *Chiral and Topological Nature of Magnetic Skyrmions*,
Springer Theses, https://doi.org/10.1007/978-3-319-98252-6_4

The topological winding number of a skyrmion structure can be experimentally inferred from the electrical transport measurements, [9–12], as well as monitoring the skyrmion Hall effect in real-space [13]. However, more decisive and straightforward experimental technique is needed to give unambiguous answer. Here, we will show that based on the physics of light-matter interaction, a direct relationship between the topological winding number of the spin texture and the polarised resonant x-ray scattering process can be established. This relationship provides a solid, one-to-one correspondence between the measured scattering signal and the winding number. We will then demonstrate that the exact topological quantities of the skyrmion material Cu_2OSeO_3 can be directly experimentally determined in this way. This universal technique has the potential to be applicable to a wide range of materials, allowing for a direct determination of their topological properties.

4.1 REXS Polarisation Dependence

We will start by writing down Eqs. (2.15) and (2.18) again here, to remind the reader of the REXS form factor and the magnetic scattering amplitude in the electric dipole-approximation for dipole transition. The scattering intensity $I(\mathbf{q})$ for the incommensurate spin lattice can be written as:

$$f^{\text{res}} = f_0(\boldsymbol{\epsilon}_s^* \cdot \boldsymbol{\epsilon}_i) - if_1(\boldsymbol{\epsilon}_s^* \times \boldsymbol{\epsilon}_i) \cdot \mathbf{m}_n \ , \tag{4.1}$$

$$\mathscr{F}_1^{\text{res}}(\mathbf{q}) = -if_1 \sum_n (\boldsymbol{\epsilon}_s^* \times \boldsymbol{\epsilon}_i) \cdot \mathbf{m}_n \, e^{i\mathbf{q} \cdot \mathbf{R}_n^{\text{Cu}}} = -iF_1 \, (\boldsymbol{\epsilon}_s^* \times \boldsymbol{\epsilon}_i) \cdot \mathbf{M}(\mathbf{q}) \ , \tag{4.2}$$

$$I(\mathbf{q}) = |\mathscr{F}_1^{\text{res}}|^2 \ . \tag{4.3}$$

Next, the detailed form of $\boldsymbol{\epsilon}_s$ and $\boldsymbol{\epsilon}_i$ will be considered. The polarisation dependence essentially gives the relationship between $I(\mathbf{q})$ and the incident light polarisation $\boldsymbol{\epsilon}_i$, for a certain magnetisation profile. As will be shown using the density matrix approach, the specific scattered polarisation of $\boldsymbol{\epsilon}_s$ does not matter in the end, as the calculated $I(\mathbf{q})$ will cover all possible $\boldsymbol{\epsilon}_s$.

The coordinate system used for carrying out the polarisation-dependent scattering cross-section is defined in Fig. 4.1b, in which the Cartesian coordinates are purely governed by the scattering plane, i.e., the xz-plane is parallel to the scattering plane, and the x-y-plane is perpendicular to the diffraction wavevector $\mathbf{Q}$. This also defines the components of the vectors, as well as the reciprocal space (q_x, q_y, q_z). Therefore, $\mathbf{k}_i = k(\cos\alpha, 0, \sin\alpha)$, $\mathbf{k}_s = k(\cos\alpha, 0, -\sin\alpha)$, $\mathbf{k}_s \times \mathbf{k}_i = k(0, -2\cos\alpha \sin\alpha, 0)$, where k is the length of the x-ray wavevector, which relates to the photon energy.

We then use the density matrix μ, written as a 2×2 matrix:

$$\mu = \frac{1}{2}(P_0\sigma_0 + \mathbf{P} \cdot \sigma) = \frac{1}{2}\begin{pmatrix} P_0 + P_1 & P_2 - iP_3 \\ P_2 + iP_3 & P_0 - P_1 \end{pmatrix} \tag{4.4}$$

where $\sigma_0 = \begin{pmatrix} 1 & 0 \\ 0 & 1 \end{pmatrix}$, and σ are the Pauli matrices, $\sigma_1 = \begin{pmatrix} 1 & 0 \\ 0 & -1 \end{pmatrix}$, $\sigma_2 = \begin{pmatrix} 0 & 1 \\ 1 & 0 \end{pmatrix}$, and $\sigma_3 = \begin{pmatrix} 0 & -i \\ i & 0 \end{pmatrix}$. $\mathbf{P} = (P_0, P_1, P_2, P_3)$ is the Poincaré-Stokes representation of light polarisation. For circularly polarised light, $P_0 = 1$, $P_1 = 0$, $P_2 = 0$, $P_3 = \pm 1$. For linearly polarised light, $P_0 = 1$, $P_1 = \cos(2\beta)$, $P_2 = \sin(2\beta)$, $P_3 = 0$.

The density matrix, $\mu_i = \sum_{\lambda_i} |\lambda_i\rangle a_{\lambda_i} \langle \lambda_i|$, specifies the polarisation state for the incident light, with the degree of polarisation $\mathscr{P}_i = \langle \sigma \rangle = \mathrm{Tr}[\sigma \cdot \mu_i]$; where $|\lambda_i\rangle$ is the diagonalised matrix for the incident polarisation eigenstates, and a_{λ_i} is the probability for that eigenstate. The polarisation state for the scattered light is then written as $\mu_s = \mathscr{F} \cdot \mu_i \cdot \mathscr{F}^\dagger$, with the degree of polarisation $\mathscr{P}_s = \mathrm{Tr}[\sigma \cdot \mu_s]$; where $\mathscr{F}$ is the operator for the scattering form factor.

On the other hand, the density matrix formalism gives rise to the form of the scattered light:

$$\mathscr{P}_s I(\mathbf{Q}) = \sum_{\lambda_i \lambda_s} a_{\lambda_i} a_{\lambda_s} \langle \lambda_i | \mathscr{F}^\dagger | \lambda_s \rangle \langle \lambda_s | \mathscr{F} | \lambda_i \rangle . \tag{4.5}$$

Therefore, the scattering cross-section with incident polarisation $\mathbf{P}$ and $\mathscr{P}_i = 100\%$, which integrates over all the possible scattered light polarisations takes the form of:

$$I(\mathbf{Q}) = \mathrm{Tr}[\mu_s] = \mathrm{Tr}[\mathscr{F} \cdot \mu_i \cdot \mathscr{F}^\dagger] . \tag{4.6}$$

Consequently, this gives rise to [14]:

$$\begin{aligned} I(\mathbf{Q}) = {} & \frac{1}{2}|F_1|^2 (P_0 + P_1)|\mathbf{k}_s \cdot \mathbf{M}(\mathbf{Q})|^2 \\ & + \frac{1}{2}|F_1|^2 (P_0 - P_1)[|\mathbf{k}_i \cdot \mathbf{M}(\mathbf{Q})|^2 + |(\mathbf{k}_s \times \mathbf{k}_i) \cdot \mathbf{M}(\mathbf{Q})|^2] \\ & - |F_1|^2 \mathrm{Re}[(P_2 + iP_3)(\mathbf{k}_s \cdot \mathbf{M}^*(\mathbf{Q}))(\mathbf{k}_s \times \mathbf{k}_i) \cdot \mathbf{M}(\mathbf{Q})] , \end{aligned} \tag{4.7}$$

in which $\mathbf{k}_i$ and $\mathbf{k}_s$ are tuned to satisfy the diffraction condition for $\mathbf{Q}$. Here we use the upper case, as it relates to the vector sum over the Bragg peak and magnetic wavevector: $\mathbf{Q} = \mathbf{G} + \mathbf{q}_m$. This is the final, and the most useful form of the polarisation dependence in our approximation, as all the parameters we care about (i.e., incident polarisation P_1, P_2, P_3, the diffraction condition $\mathbf{k}_i$ and $\mathbf{k}_s$, as well as the magnetisation profile $\mathbf{M}(\mathbf{Q})$) are explicitly expressed in Eq. (4.7). Moreover, the quantities that we do not care about (such as the outgoing polarisation $\mathbf{P}_s$ or $\mathscr{P}_s$, the Bragg peaks $\mathbf{G}$, as well as the polarisation dependence results from the charge parts) are automatically dealt with, and do not appear in the polarisation dependence relationship. $|F_1|$ is treated as a constant as it only relates to k, and only spectroscopists would specifically worry about it.

Next, it is worth highlighting that this equation is vector-sensitive, i.e., polarised x-rays probe all three components of the magnetisation vector. As can be seen from the scattering coordinates, the $\mathbf{k} \cdot \mathbf{M}(\mathbf{Q})$ terms are only sensitive to the Fourier transforms of m_1 and m_3 components, while the $(\mathbf{k}_s \times \mathbf{k}_i) \cdot \mathbf{M}(\mathbf{Q})$ term encodes the information of m_2. For example, if the incident light is σ-polarised, Eq. (4.7) reduces to Eq. (2.24), in which the magnetic diffraction signal at $\mathbf{Q}$ only reflects the Fourier transform of m_1 and m_3. At this point, if the experimentalist can orient the sample such that the magnetic moments have the maximum number pointing towards m_2, minimum diffraction intensity is observed. Similarly, if one can scan the same $\mathbf{Q}$ for its different azimuthal positions, a systematic diffraction intensity as a function of the azimuthal angle can be measured, giving the information about how m_2 arranges inside of the motif. Most interestingly, the phase relation that happens for circular polarisation, expressed by the iP_3 term, can be very powerful revealing the chiral information of the spin structure [14].

From this, we hope it becomes clear that magnetic diffraction is fundamentally different from charge diffraction due to the vector nature of the magnetisation. For charge diffraction from the atomic lattice, the motifs are simply spherically symmetric atoms. The form factor of the motifs thus contains only scalars, which are isotropic, i.e., for a Bragg peak, the diffraction intensity has no azimuthal dependence. In other words, the x-rays will 'see' a charge order in the same way from any direction. However, if the motif is composed of magnetic moments, the x-rays will 'see' the same motif differently from different directions. (Think about an object, and you look at its shape when you go around it: if this object is a ball, all views are indistinguishable; however, if this object is an arrow, you will see it differently at different angles.) This provides the key idea for the study of the motif structure at a magnetic diffraction condition: we try to let the x-rays 'see' the structure from different angles (at the same diffraction peak), and guess the structure of the motif from these angular dependence relationships. This is achieved by changing the relative angle between the light polarisation and the spin vectors. Consequently, the experiment can be carried out in two ways: (1) rotate the spins; (2) rotate the light polarisation. Here, we will introduce some methodologies that are specifically suitable for long-wavelength modulated chiral spin structures.

4.2   Topology Determination Principle

In a many-body system, an order parameter can be assigned to the individual entities, and by considering interactions among them, emergent phases and novel physical properties may evolve [15]. The possible values of the order parameter constitute the order parameter space, which can be described in the framework of topology [16]. In magnetism, the spins are the elementary entities, and the order parameter is the magnetisation vector $\mathbf{m}$. Its magnitude can be taken as a constant, i.e., its three components satisfy $m_1^2 + m_2^2 + m_3^2 = M_S^2$, where M_S is the saturation magnetisation. Therefore, the order parameter space is the surface of a three-dimensional sphere, which is described by the homotopy group $\pi_2(S^2)$ for a two-dimensional physical space (x, y) [7], as discussed in Chap. 1. Different homotopy classes with distinct topological properties can be quantified based on the winding number N, which is an integer that counts the number of times the physical space wraps around the order parameter space [6]. It is defined by Eq. (1.10). If the two-dimensional real-space lies in x-y-plane, N takes the form of:

$$N = \frac{1}{4\pi} \int \left(\frac{\partial \mathbf{m}}{\partial x} \times \frac{\partial \mathbf{m}}{\partial y} \right) \cdot \mathbf{m} \, dx \, dy \; . \tag{4.8}$$

It has been demonstrated that helimagnetic materials with homotopy class $N = 1$ [8], i.e., the magnetic skyrmions [17], induce emergent electromagnetism [10, 18], giving rise to a number of novel magnetoelectric phenomena [9, 10, 19]. Utilising this nontrivial topological order, advanced spintronics applications have been devised [18, 20–23]. More recently, several candidates with $N = 2$ have also been discovered [24, 25], suggesting that other elements from the $\pi_2(S^2)$ group may exist in nature as well.

While the significance of the topological properties of ordered systems is being recognised more and more, the experimental determination of the winding number for spin-ordered media remains challenging. Commonly, the winding number is determined by comparing a microscopic image of the magnetisation state with theoretical model calculations, making it a rather indirect process that has no unique answer [20–23, 26–28]. Most importantly, the established magnetic imaging techniques only give a partial picture of the local magnetisation vector, as they are both limited in three-dimensional sensitivity and lateral resolution. Electric transport measurements, on the other hand, despite being able to reveal details of the local magnetic structure [9], are usually also affected by other, non-topological contributions to the transport [29], rendering them less ideal for the unambiguous determination of topological properties. Here, we will show that the winding number N can be unambiguously measured using x-rays, utilising the sensitivity of the light polarisation to the magnetic order, called the *topology determination principle* (TDP).

The skyrmion solution with a topological winding number N is given by Eq. (1.11). However, another more illustrative way to construct a topological entity is to wrap a one-dimensional proper-screw helical pitch around a centre axis N times, where N is the topological winding number. Note that the skyrmion structure has to be axial-symmetric in this way. This construction is illustrated in Fig. 4.1a. The exact type of the underlying helix does not affect the topological properties, as will be shown below. In this way, we can easily obtain an analytical solution which is the basis for the TDP. As shown in Fig. 4.1b, the traversed physical space is described by the azimuthal angle Ψ, where $\Psi = 0°$ corresponds to the x-direction. As the helical pitch wraps around a circle once, when Ψ changes by $360°$, the full order parameter space is covered, which gives $N = 1$. As illustrated in Fig. 4.2a–d, also motifs with different topological winding numbers can be constructed this way, which is equivalent to the description given in Eq. (1.11). In other words, if the physical space traverses Ψ by rotating the sample, the helix would traverse $N\Psi$ in order to cover the order parameter space.

The TDP measurement geometry is illustrated in Fig. 4.1b. The incident and scattered x-ray wavevectors are denoted as $\mathbf{k}_i$ and $\mathbf{k}_s$, with the incident angle α, which satisfies the diffraction condition $\mathbf{Q} = \mathbf{k}_s - \mathbf{k}_i$. The incident x-rays can be

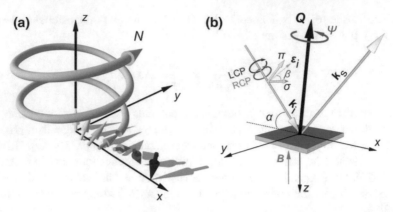

Fig. 4.1 Concept and experimental setup for determining the topological winding number using polarised x-rays. **a** For classical spins in two-dimensional space, the spin configuration that carries a winding number of N, as described by Eq. (1.11), can be equivalently constructed by encircling a one-dimensional proper-screw-type spin helix pitch around a centre N times. **b** When the diffractive x-rays probe the entire physical space by rotating the topological motif within the xy-plane by 360°, the scattered intensity will exhibit a periodicity that is only dependent on N. Either circularly or linearly polarised incident light can be used. In both cases, the diffraction condition is met for the wavevector $\mathbf{Q}$, which contains the topological motif's modulation wavevector, for different azimuthal angles Ψ

linearly polarised with the polarisation angle β. We define $\beta = 0°$ corresponding to σ-polarisation, while $\beta = 90°$ corresponds to π-polarisation. Alternatively, the light can be circularly polarised. Here, we define the circular dichroism (CD) signal as the difference of the scattering cross-sections for left-circularly and right-circularly polarised incident light (at the same diffraction condition).

We demonstrate the TDP principle on the familiar magnetic skyrmion system Cu_2OSeO_3. As we have studied before, the modulation wavevector is ~0.0158 (r.l.u.), and the motif's periodic lattice lies in the xy-plane when the required magnetic field is along z-direction [30], as shown in Fig. 4.1. Followed by the 'winding-one-dimensional-helix' concept, the skyrmion vortex is equivalent to wrapping the helix once while the physical space traverses 360°.

Therefore, a one-dimensional proper-screw helix pitch [31], otherwise called 'Bloch-type' helix, propagating along x is written as:

$$
\begin{aligned}
m_1 &= 0 \,, \\
m_2 &= M_S \cos(\mathbf{q}_h \cdot \mathbf{r}) \,, \\
m_3 &= M_S \sin(\mathbf{q}_h \cdot \mathbf{r}) \,.
\end{aligned}
\tag{4.9}
$$

where $\mathbf{q}_h$ is the helix propagation wavevector, $\mathbf{r} = (x, y)$. This is defined as $\Psi = 0°$ case. While the x-rays probe in physical space at Ψ angle, the propagation wavevector would rotate $N\Psi$ within the q_x-q_y plane, forming a modulation structure of:

$$m_1 = -M_S \sin(N\Psi) \cos(\mathbf{q}_h \cdot \mathbf{r}) \ ,$$
$$m_2 = M_S \cos(N\Psi) \cos(\mathbf{q}_h \cdot \mathbf{r}) \ , \tag{4.10}$$
$$m_3 = M_S \sin(\mathbf{q}_h \cdot \mathbf{r}) \ .$$

This is nothing but applying the rotation matrix $\begin{pmatrix} \cos(N\Psi) & -\sin(N\Psi) & 0 \\ \sin(N\Psi) & \cos(N\Psi) & 0 \\ 0 & 0 & 1 \end{pmatrix}$ on the magnetic structure in Eq. (4.9). To meet the diffraction condition for $\mathbf{Q} = \mathbf{G} + \mathbf{q}_h$ at Ψ, one has to bring $\mathbf{Q}$ into the scattering plane. For a common four-circle diffractometer, this is achieved by compensating the diffraction offset with the other two axis, i.e., $\mathscr{X}$ axis that is perpendicular to both Ω axis and Ψ axis; as well as Ω axis that gives rise to α. Consequently, the coordinates that decomposing the magnetic structures transform into:

$$\begin{pmatrix} m'_1 \\ m'_2 \\ m'_3 \end{pmatrix} = \mathscr{R}_\Omega \mathscr{R}_{\mathscr{X}} \begin{pmatrix} m_1 \\ m_2 \\ m_3 \end{pmatrix} \ . \tag{4.11}$$

where $\mathscr{R}_{\mathscr{X}}$ and $\mathscr{R}_\Omega$ are the rotation matrix, and the combination of the two rotations brings $\mathbf{Q}$ into the scattering plane for the diffraction condition. However, it is essential to note that this change would be negligible for most of the long-wavelength modulated magnetic structures. For example, Cu_2OSeO_3 has $q/G = 0.0158$ [30], where $G = 1$, corresponds to the (0,0,1) peak. Therefore, the change of Ω is less than 0.9° for all Ψ angles. This makes $\mathscr{R}_{\mathscr{X}}$ and $\mathscr{R}_\Omega$ negligible as well. Therefore, the long-wave-length approximation suggests $\begin{pmatrix} m'_1 \\ m'_2 \\ m'_3 \end{pmatrix} \approx \begin{pmatrix} m_1 \\ m_2 \\ m_3 \end{pmatrix}$. As a result, $\mathbf{M}(\mathbf{Q}) \approx \mathbf{M}(\mathbf{q}_h)$. Furthermore, one should keep one α angle for the diffraction condition for all Ψ positions.

Thus, the Fourier transform of Eq. (4.10) at the diffraction condition of $\mathbf{q}_h$ takes the form of:

$$M_1(\mathbf{q}_h, \Psi) = -\pi M_S \sin(N\Psi) \ ,$$
$$M_2(\mathbf{q}_h, \Psi) = \pi M_S \cos(N\Psi) \ , \tag{4.12}$$
$$M_2(\mathbf{q}_h, \Psi) = -i\pi M_S \ .$$

Inserting Eq. (4.12) into Eq. (4.7), and evaluating the expressions described above, the CD profile is written as:

$$I_{CD} = 4Y \sin^2\alpha \, \cos\alpha \, \cos(N\Psi) \ . \tag{4.13}$$

Also, the linear polarisation dependence, called polarisation-azimuthal-map (PAM), can be derived as:

$$I = Y\sin^2\alpha + Y\cos^2\alpha \, \sin^2(N\Psi) + 4Y\cos^2\alpha \, \sin^2\alpha \, \cos^2(N\Psi) \sin^2\beta$$
$$- Y\cos^2\alpha \, \sin\alpha \, \sin(2N\Psi) \sin(2\beta) \; . \tag{4.14}$$

in which $Y = \pi^2|F_1|^2k^2M_S^2$. Note that the CD cross section extinct for even number of N. This feature cannot be captured using the analytical relationship of Eq. (4.13), however can be generalised from the numerical calculations. The reason why CD vanishes is due to the neutralised total chirality of the even winding number spin configuration. This can be exemplified by extracting an arbitrary one-dimensional spin chain from the two-dimensional configuration of the motif, such as in Fig. 4.2b. As can be seen from this chain, the spins first wind in one sense towards the centre, then unwind back with the opposite rotating sense. Therefore, the chirality is cancelled out, giving rise to no CD. This chirality-cancelling effect does not occur for odd winding number motifs, however exist for all even winding numbers.

Equations (4.13) and (4.14) are the basic form of TDP, which are derived based on the 'encircling-proper-screw-helix' concept introduced in Fig. 4.1a. Therefore the analytical form is only valid for the 'standard' spin configuration. However, in principle there are infinite homotopies for the $N = 1$ class, i.e., the same topological property will always hold if continuous transformations are operated on the 'standard' motif as shown in Fig. 4.1a. For example, if the encircling one-dimensional helix takes other form, such as the cycloidal-type [25], the overall spin texture will change, while the winding number keeps invariant. As will be proven later, This 'homotopy' parameter is dealt with adding a phase factor into Eqs. (4.13) and (4.14), which makes TDP valid for all cases.

Adding all these corrections into the CD and PAM profiles, the analytical form that specifies the direct relationship between of the resonant magnetic diffraction cross-section and N is summarised as:

$$I(\beta, \Psi, N) = Y \sin^2\alpha + Y \cos^2\alpha \, \sin^2(N\Psi + \Phi_2)$$
$$+ 4Y \cos^2\alpha \, \sin^2\alpha \, \cos^2(N\Psi + \Phi_2) \sin^2\beta \tag{4.15}$$
$$- Y \cos^2\alpha \, \sin\alpha \, \sin(2N\Psi + 2\Phi_2) \sin 2\beta \, ,$$

and

$$I_{CD}(\Psi, N) = \begin{cases} \mathscr{C} \cos(N\Psi + \Phi_1) \; ; \; \text{for } N \text{ is odd}, \\ 0 \qquad\qquad\quad ; \; \text{for } N \text{ is even}, \end{cases} \tag{4.16}$$

where $\mathscr{C}$ and Y are constants. The arbitrary phase parameters Φ_1 and Φ_2 adapt the continuous transformations onto the local spin structure, i.e., the spin configuration that has the same winding number, however, which deviates from the 'standard' configuration as shown in Fig. 4.1a. These two relationships can be interpreted in the following way: *In case of an odd winding number, N equals to the periodicity of the CD signal while the x-rays probe the full physical space once. N is also equal to half the number of peaks in the polarisation-azimuthal map (see below). In case*

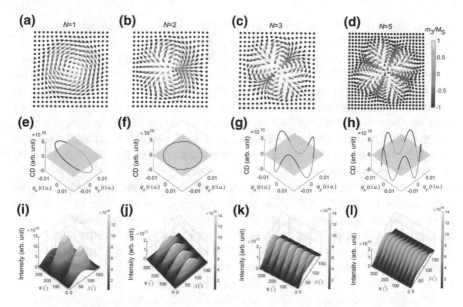

Fig. 4.2 Calculated results demonstrating the topology determination principle. **a–d** Spin configurations with topological winding numbers of 1, 2, 3, and 5, respectively. They can be either constructed using Eq. 1.11), or using the concept illustrated in Fig. 4.1a. These spin textures are the motifs that are repeated in the two-dimensional, periodically ordered structures which can be measured by diffraction techniques. **e–textbfh** CD cross-section as a function of Ψ for topologically ordered systems. Here, the Ψ parameter is transformed into reciprocal space (q_x, q_y) for a constant length of $q = 0.0158$ (r.l.u.) (in the coordinate system of Cu_2OSeO_3). The CD extinction condition (CD = 0) is indicated by the light blue plane. The CD signal is symmetric about this plane for integer winding numbers. The periodicity of the CD modulation equals to N. Note that for even N the CD signal is zero. **i–l** Polarisation-azimuthal maps. The calculations are performed by rotating β from $0°$ to $180°$ at each Ψ, and by mapping out Ψ from $0°$ to $360°$. The total number of the humps is equivalent to $2N$. Note that for integer winding numbers the humps are of equal height

of an even winding number, no CD signal is observed. This is called the topology determination principle.

Figure 4.2e–h shows the numerical calculation of the CD cross-section for different topological spin motifs. Each reciprocal space point (q_x, q_y) on the red closure lines corresponds to one azimuthal angle of the sample, at which the diffraction condition for the modulation wavevector propagating at Ψ is met. Therefore, according to Eq. (4.16), the CD intensity modulates while circling around the physical space, with a periodicity that equals to N, if the spin winding is an odd number. For even winding numbers, such as $N = 2$, there is no CD. This is because the superposition of the equal number of 'winding' and 'unwinding' instances does not possess global chirality. On the other hand, using linearly polarised light, one can measure the polarisation-dependent scattering intensity at each Ψ. By carrying out this measurement for the full $360°$ physical space, a polarisation-azimuthal map can be obtained. The PAM plot in Fig. 4.2i–l shows hump-like, two-dimensional peaks of equal height. The peaks appear around $\beta \approx 90°$, and modulate along the physical space. Their periodicity is twice that of the winding number. This PAM feature is well described analytically by Eq. (4.15). If both the CD and PAM measurements can be fitted by Eqs. (4.15) and (4.16) at the same time, the winding number can be unambiguously determined by the TDP. As will be shown below, if the detailed spin structure of the motif modifies, such that the exact mapping from physical space to order parameter space changes within the same homotopy class, the corresponding shapes of the CD and PAM only undergo a linear shift, while the periodicities do not change. Therefore, the topological robustness is also reflected in this type of x-ray scattering measurement.

We performed the TDP experiments on single-crystalline Cu_2OSeO_3, with the setup sketched in Fig. 4.1b. At 57 K, and in an applied magnetic field of 32 mT, the skyrmion lattice phase emerges, manifesting itself as a hexagonal lattice of $N = 1$ topological motifs. The lattice gives rise to the six-fold-symmetric diffraction pattern in reciprocal space, shown in Fig. 4.3a. The sharp six first-order magnetic peaks correspond to the 'unit cell' of the skyrmion lattice, with one of them locked along h, i.e., along the [100] crystallographic direction in real space. This is due to the higher-order magnetic anisotropy [30, 32]. Note that the coordinates (q_x, q_y) used here, as defined in Fig. 4.1b, are independent of the crystallographic directions. Therefore, by rotating Ψ, the same modulation wavevectors rotate accordingly in our coordinate system (see Fig. 4.3b and c). The measured CD intensity as a function of Ψ (cf. Fig. 4.3d) shows exactly one period for the x-rays mapping the physical space once, suggesting that the skyrmion motif has a topology of $N = 1$. Moreover, as shown in Fig. 4.3e and f, the PAM is in excellent agreement with the theoretical calculations shown in Fig. 4.2i. The two equal humps confirm once more the $N = 1$ topological property of this material.

Another type of $N = 1$ system, which does not carry a net chirality, is the so-called Néel-type skyrmion [33] (see Fig. 4.4a). Its spin texture has a different appearance, however, it is topologically equivalent to the other skyrmion form. Consequently, the CD profile in Fig. 4.4b shows the same periodicity, however, a constant phase shift, as compared with Fig. 4.2e. At first sight, as a Néel-type skyrmion shows no net chirality, it is puzzling why it results in a CD signal. Its appearance can be explained by the concept illustrated in Fig. 4.1a. In case of the Néel-type skyrmion the one-dimensional basis pattern is a cycloidal-type modulation, in which the spins rotate in a common plane that is parallel to its propagation direction. The chirality, however, remains encoded in this common plane relative to the helix propagation, which can be effectively picked up by circularly polarised x-rays [34]. The phase shift, on the other hand, is due to the different mapping of the spin configuration under a continuous transformation. The same behaviour is found for the PAM, as shown in Fig. 4.4e, for which the shape and height of the two humps are essentially the same as in Fig. 4.2i, however, the entire pattern undergoes a linear shift along Ψ. Moreover, if the winding number takes a negative value (cf. Fig. 4.4d) the CD still shows the same periodicity, i.e., it does not distinguish between N and $-N$. However, the shape of the humps is fundamentally different (compare Fig. 4.4f with Fig. 4.4c and 4.2i).

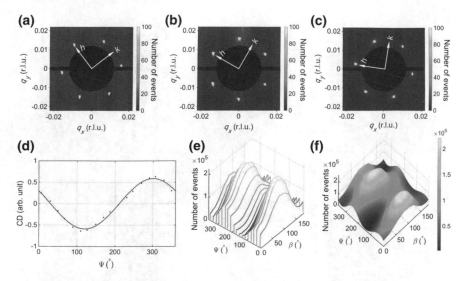

Fig. 4.3 Experimental results of the winding number measurement in the skyrmion lattice phase of Cu_2OSeO_3. **a–c** Resonant soft x-ray magnetic diffraction of Cu_2OSeO_3. The photon energy is tuned to 931.25 eV and the magnetic satellites are collected around the (0, 0, 1) structural peak. The figure shows the reciprocal space maps within the hk-plane at $l = 1$. The coordinate system is defined in Fig. 4.1b. The temperature is 57 K, and the applied magnetic field is 32 mT along z-direction, which is also parallel to the [001] crystallographic orientation. **a–c** corresponds to three different azimuthal angles, with $\Psi = 5.95°$, $29.56°$, and $50.17°$, respectively. **d** CD signal as a function of Ψ. The black dots are the measured data, while the red line is a fit using Eq. (4.16). **e** Measured PAM, and **f** interpolation obtained by fitting Eq. (4.15). The interpolated PAM shows excellent agreement with the calculated result shown in Fig. 4.2i, in which two humps of equal height can be observed at $\beta \approx 90°$

In summary, we have demonstrated that for an incommensurate long-wavelength magnetically ordered system, the topological winding number of the motif can be unambiguously determined by polarised x-rays. First, the TDP is a direct measurement method as the winding number is naturally encoded in the underlying physics of the light-matter interaction. Second, although we used resonant soft x-ray diffraction for the demonstration of the measurement principle, the same theory can be applied to the hard x-ray region as well, and it can be, in principle, expanded to non-resonant magnetic x-ray scattering. Third, the TDP can be applied to a wide range of materials that carry topologically ordered spins, making TDP a general experimental principle.

4.3 Robustness of Topology Determination Principle

So far, we have demonstrated the excellent consistency of the analytical solution, numerical calculation and experimental data for the TDP. Here, we will prove the robustness of this principle.

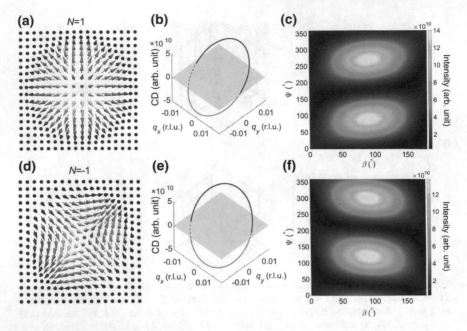

Fig. 4.4 Robustness and validity of the topology determination principle. **a** Spin configuration of a Néel-type skyrmion with $N = 1$. **b** Calculated CD profile, and, **c** calculated PAM for the Néel-type skyrmion. **d** Spin configuration for an antiskyrmion with $N = -1$. Nevertheless, the TDP can be applied, as shown in (**e**–**f**), in which the CD profile and PAM periodicity suggest the correct absolute value of the winding number. Moreover, negative winding numbers give rise to a fundamentally different PAM shape of the humps. This provides an additional way to distinguish positive from negative N

4.3.1   Numerical Calculations and Robustness of TDP

The numerical calculations are carried out based on the materials parameters of Cu_2OSeO_3, using a helix pitch of 60 nm. This leads to a skyrmion core-to-core distance of $\sim$69.28 nm, as well as the wavevector of $\sim$0.015 (r.l.u.). Resonant x-ray scattering at the Cu L_3 edge with a photon energy of 931.25 eV gives $k = 2\pi \times 0.751 \, nm^{-1}$, with $\alpha \approx 48.24°$ for the (0,0,1) diffraction peak. In the calculation, F_1 and M_S are kept constant as the CD profile and PAM are measured for the same photon energy and temperature.

Figure 4.5a shows as an example the cropped simulation object with $N = 1$. The motifs are generated based on Eq. (1.11) and assemble into a hexagonally ordered two-dimensional periodic lattice. The symmetry of the lattice does not change any properties of the results, as we only concentrate on one wavevector at different Ψ. In the calculations we use a size of $300 \times 300 \, nm^2$ of the periodic motif lattice. The calculated diffraction intensity, shown as reciprocal space maps using Eq. (4.7), are plotted in Fig. 4.5b and c, where the incoming light is at $\beta = 0°$ and $90°$, respectively. On the other hand, we use the 'encircling-one-dimensional-helix' method to perform

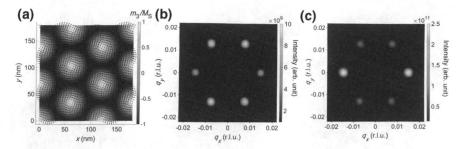

Fig. 4.5 Numerical simulation object. **a** Cropped real-space periodic lattice of the topological motif. The figure shows a hexagonally ordered array of $N = 1$ motifs, i.e., the skyrmion lattice as observed in Cu_2OSeO_3. Note that, in principle, the motif ordering can take any symmetry, such as two-fold, three-fold, or four-fold, which always gives rise to the same TDP results. **b, c** Calculated magnetic diffraction pattern in reciprocal space, based on the real-space spin configuration in (**a**). The material's parameters used in the calculation are based on Cu_2OSeO_3. The photon energy is 931.25 eV, with σ-polarisation for (**b**), and π-polarisation for (**c**)

the numerical calculations for the same object, in order to confirm the equivalence of both methods.

We first prove the consistency between both calculation methods, which further confirms the validity of our analytical solution. Figure 4.6 shows the comparison for PAMs obtained using the two different methods. The top and bottom row correspond to $N = 1$ and $N = 3$, respectively. Plots in the left column arise directly from Eq. (4.15); plots in the middle column are the numerical calculation results using the 'encircling-one-dimensional-helix' method; while those in the right column are the numerical results based on the spin configurations defined by Eq. (1.11). As can be seen, the general feature of the multiple humps is consistent, which is directly linked to the winding number. The deviations from each other only lie in some fine details, which are due to the exact values of the parameters used in the numerical calculations, such as M_S, k, and α. Also, we found that the real-space boundaries, and the simulation mesh structure can cause slight deviations compared with Eq. (4.15), which can be regarded as simulation artefacts. This consistency demonstrates that the concept of the one-dimensional-spin helix indeed allows us to construct a topological spin structure with winding number N and that the analytical solution is valid.

Next, we show that other parameters in Eq. (1.11), including the $\theta(\rho)$ function, λ, and χ do not affect the validity of the TDP. First, we discuss the dependence on $\theta(\rho)$, which can be regarded as the radial profile of the out-of-plane magnetisation component, m_3, from the core to the boundary. For an axially symmetric entity, $\theta(\rho)$ usually satisfies the Euler-Lagrange equation [35]. However, for our purpose, changing this profile does not affect the CD and PAM features at all, as long as the boundary condition is met. We use the $N = 1$ skyrmion as an example. As shown in Fig. 4.7a, $\rho = 0$ nm corresponds to the real-space skyrmion vortex core, while the maximum value of ρ in the plot corresponds to the radius of a skyrmion disk in our calculations. Its value is related to the helix pitch by $(60 \text{ nm} \times 2/\sqrt{3})/2$.

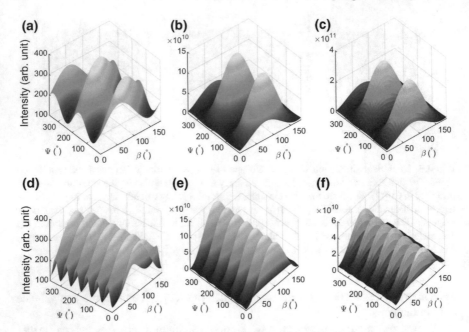

Fig. 4.6 Comparison between different TDP simulation methods. **a–c** PAM for the $N = 1$ motif and, **d–f**, for the $N = 3$ motif calculated using three different methods: **a, d** are obtained directly using Eq. (4.15); **b, e** are calculated based on the diffraction condition for traversing a one-dimensional helix $N\Psi$ around a centre at different angles Ψ; **c, f** are calculated based on a rigorous two-dimensional periodic lattice formed by topological motifs with $N = 1$ and $N = 3$, respectively

Three different $\theta(\rho)$ profiles, labelled as (i), (ii), and (iii), are used in the simulation for the purpose of comparison. All of them have the same homotopy. As shown in Fig. 4.6b–g for both CD and PAM, the three different profiles do not induce any significant differences. They are nearly identical, except for the slight difference in amplitude of the cross-section. This is to be expected as the TDP specifies the measurement under diffraction conditions, therefore the detailed modulation profile along the propagation direction is only reflected in the total scattering amplitude, not the polarisation dependence.

Second, we discuss the influence of χ and λ on the CD and PAM pattern. Figure 4.8a shows another $N = 3$ topological object, which is essentially a continuous transformation from the object in Fig. 4.2c. This homotopic transformation can be achieved by adjusting χ. As shown in Fig. 4.8b and c, compared to Fig. 4.2g and k, the CD and PAM patterns take the identical periodicities, and the only difference is a linear phase shift. This is valid for all cases in our numerical studies. Moreover, as shown in Fig. 4.8d–f, flipping the polarity of the topological object does not alter PAM, however, it imposes a phase shift of the CD profile. Therefore, the use of the phase parameters Φ_1 and Φ_2 in Eqs. (4.15) and (4.16) can generalise the principle to all homotopies arising from variations in χ and λ.

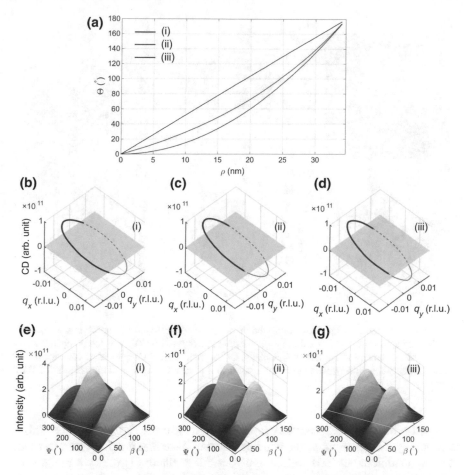

Fig. 4.7 Robustness of the TDP for different homotopies generated by changing the radial profile. **a** Three different $\theta(\rho)$ profiles that govern different radial spin distributions are used for the subsequent numerical calculations, labelled as (i), (ii), and (iii). Note that profile (i) represents a linear relationship, which is equivalent to the one-dimensional helix modulation case. **b–d** CD profiles and, **e–g** PAM calculated based on the three different radial functions

Briefly summarised, the TDP, represented by the two polarisation-dependent profiles CD and PAM, is only sensitive to the winding number, and has a one-to-one correspondence to this topological quantity. Any homotopy change will not affect its validity. In other words, the TDP itself is 'topologically protected'.

Up to here, we have presented a detailed treatment of the REXS polarisation dependence which is an explicit function of the topological winding number. The reason why TDP is a direct method to measure N can be explained this way (of course one can refer to Eqs. (4.15) and (4.16), and have a taste of how 'direct' they are): the azimuthal scan plays the role of the physical space, while the REXS intensity counts

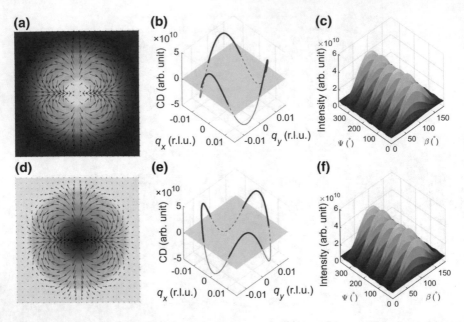

Fig. 4.8 Robustness of the TDP for different homotopies generated by modifying χ and λ. **a** Real-space spin configuration with $N = 3$, however, having a different χ value compared to the configuration shown in Fig. 4.2c. **b** Calculated CD profile, and **c** PAM pattern. **d** Another $N = 3$ spin configuration with opposite polarity λ compared to **a**. **e** Calculated CD profile, and **f** PAM pattern

how often the physical space has wrapped around the order parameter space. This is essentially the definition of the winding number. Note that it is important to check the preconditions for TDP before applying this method: (1) the spin modulation has to have a harmonic eigenmode; (2) the periodicity of the modulation has to be incommensurate, and has to fall into the long-wavelength approximation. Fortunately the validity of the two can be confirmed from normal magnetic diffraction peaks: the magnetic peak positions will reflect the periodicity of the modulation, and directly tell whether they are commensurate or not. Furthermore, some simple azimuthal scans can confirm whether the motif is harmonic or not. Also the non-zero dichroism can further confirm that the preconditions are met.

At this point the reader may have noticed that there is a loose end in those two equations, which we have not yet dealt with carefully, i.e., the 'random' phases Φ_1 and Φ_2. In the following chapter, we will show that these phase factors are not random at all, however revealing another degree of freedom of the modulated spin motifs: the helicity angle χ.

References

1. L.H. Kauffman, *Knots and Physics* (World Scientific Publishing, 2001)
2. P.W. Anderson, G. Toulouse, Phys. Rev. Lett. **38**, 508 (1977)
3. A.M. Sonnet, E.G. Virga, *Dissipative Ordered Fluids—Theories for Liquid Crystals* (Springer, 2012)
4. J.F. Annett, *Superconductivity, Superfluids and Condensates* (Oxford University Press, 2004)
5. M.Z. Hasan, C.L. Kane, Rev. Mod. Phys. **82**, 3045 (2010)
6. H.-B. Braun, Adv. Phys. **61**, 1 (2012)
7. N.D. Mermin, Rev. Mod. Phys. **51**, 591 (1979)
8. N. Nagaosa, Y. Tokura, Nat. Nanotech. **8**, 899 (2013)
9. A. Neubauer, C. Pfleiderer, B. Binz, A. Rosch, R. Ritz, P.G. Niklowitz, P. Böni, Phys. Rev. Lett. **102**, 186602 (2009)
10. T. Schulz, R. Ritz, A. Bauer, M. Halder, M. Wagner, C. Franz, C. Pfleiderer, K. Everschor, M. Garst, A. Rosch, Nat. Phys. **8**, 301 (2012)
11. R. Ritz, M. Halder, M. Wagner, C. Franz, A. Bauer, C. Pfleiderer, Nature **497**, 231 (2013)
12. M. Lee, W. Kang, Y. Onose, Y. Tokura, N.P. Ong, Phys. Rev. Lett. **102**, 186601 (2009)
13. W. Jiang, X. Zhang, G. Yu, W. Zhang, X. Wang, M.B. Jungfleisch, J.E. Pearson, X. Cheng, O. Heinonen, K.L. Wang et al., Nat. Phys. **13**, 162 (2017)
14. G. van der Laan, C. R. Phys. **9**, 570 (2008)
15. P.M. Chaikin, T.C. Lubensky, *Principles of Condensed Matter Physics* (Cambridge University Press, 1995)
16. M. Nakahara, *Geometry, Topology and Physics* (IOP Publishing, 2003)
17. A.N. Bogdanov, D.A. Yablonskii, Sov. Phys. JETP **68**, 1 (1989)
18. N. Nagaosa, Y. Tokura, Phys. Scr. **T146**, 014020 (2012)
19. F. Jonietz, S. Mühlbauer, C. Pfleiderer, A. Neubauer, W. Münzer, A. Bauer, T. Adams, R. Georgii, P. Böni, R.A. Duine, K. Everschor, M. Garst, A. Rosch, Science **330**, 1648 (2010)
20. F. Büttner, C. Moutafis, M. Schneider, B. Krüger, C.M. Gunther, J. Geilhufe, C.V. Korff Schmising, J. Mohanty, B. Pfau, S. Schaffert, A. Bisig, M. Foerster, T. Schulz, C.A.F. Vaz, J.H. Franken, H.J.M. Swagten, M. Kläui, S. Eisebitt, Nat. Phys. **11**, 225 (2015)
21. W. Jiang, P. Upadhyaya, W. Zhang, G. Yu, M.B. Jungfleisch, F.Y. Fradin, J.E. Pearson, Y. Tserkovnyak, K.L. Wang, O. Heinonen, S.G.E. te Velthuis, A. Hoffmann, Science **349**, 283 (2015)
22. O. Boulle, J. Vogel, H. Yang, S. Pizzini, D. de Souza Chaves, A. Locatelli, T.O. Mentes, A. Sala, L.D. Buda-Prejbeanu, O. Klein, M. Belmeguenai, Y. Roussigné, A. Stashkevich, S.M. Chérif, L. Aballe, M. Foerster, M. Chshiev, S. Auffret, I.M. Miron, G. Gaudin, Nat. Nanotech. **11**, 449 (2016)
23. S. Woo, K. Litzius, B. Krüger, M.-Y. Im, L. Caretta, K. Richter, M. Mann, A. Krone, R.M. Reeve, M. Weigand, P. Agrawal, I. Lemesh, M.-A. Mawass, P. Fischer, M. Kläui, G.S.D. Beach, Nat. Mater. **15**, 501 (2016)
24. X.Z. Yu, Y. Tokunaga, Y. Kaneko, W.Z. Zhang, K. Kimoto, Y. Matsui, Y. Taguchi, Y. Tokura, Nat. Commun. **5**, 3198 (2014)
25. W. Wang, Y. Zhang, G. Xu, L. Peng, B. Ding, Y. Wang, Z. Hou, X. Zhang, X. Li, E. Liu, S. Wang, J. Cai, F. Wang, J. Li, F. Hu, G. Wu, B. Shen, X.-X. Zhang, Adv. Mater. **28**, 6887 (2016)
26. X.Z. Yu, Y. Onose, N. Kanazawa, J.H. Park, J.H. Han, Y. Matsui, N. Nagaosa, Y. Tokura, Nature **465**, 901 (2010)
27. S. Heinze, K. von Bergmann, M. Menzel, J. Brede, A. Kubetzka, R. Wiesendanger, G. Bihlmayer, S. Blügel, Nat. Phys. **7**, 713 (2011)
28. P. Milde, D. Köhler, J. Seidel, L.M. Eng, A. Bauer, A. Chacon, J. Kindervater, S. Mühlbauer, C. Pfleiderer, S. Buhrandt, C. Schütte, A. Rosch, Science **340**, 1076 (2013)
29. S.A. Meynell, M.N. Wilson, J.C. Loudon, A. Spitzig, F.N. Rybakov, M.B. Johnson, T.L. Monchesky, Phys. Rev. B **90**, 224419 (2014)
30. S.L. Zhang, A. Bauer, D.M. Burn, P. Milde, E. Neuber, L.M. Eng, H. Berger, C. Pfleiderer, G. van der Laan, T. Hesjedal, Nano Lett. **16**, 3285 (2016)

31. Y. Tokura, S. Seki, Adv. Mater. **22**, 1554 (2010)
32. T. Adams, A. Chacon, M. Wagner, A. Bauer, G. Brandl, B. Pedersen, H. Berger, P. Lemmens, C. Pfleiderer, Phys. Rev. Lett. **108**, 237204 (2012)
33. I. Kézsmárki, S. Bordács, P. Milde, E. Neuber, L.M. Eng, J.S. White, H.M. Rønnow, C.D. Dewhurst, M. Mochizuki, K. Yanai, H. Nakamura, D. Ehlers, V. Tsurkan, A. Loidl, Nat. Mater. **14**, 1116 (2015)
34. H.C. Walker, F. Fabrizi, L. Paolasini, F. de Bergevin, D. Prabhakaran, A.T. Boothroyd, D.F. McMorrow, Phys. Rev. B **88**, 214415 (2013)
35. A.N. Bogdanov, A. Hubert, J. Magn. Magn. Mater. **138**, 255 (1994)

Chapter 5
Measurement of the Skyrmion Helicity Angle

In this chapter, the measurement of the skyrmion helicity angle χ by the REXS-based technique is presented. Before starting, we would like to clarify the nomenclature used here. For a $N = 1$ skyrmion texture, as defined in Eq. (1.11), the chirality, denoted as $\mathscr{C}$ (i.e., $\mathscr{C} = \pm 1$), describes the vortex handedness. Here, $\mathscr{C} = 1$ corresponds to the right-handed skyrmion, while $\mathscr{C} = -1$ corresponds to a left-handed one. Helicity, on the other hand, denoted as $\mathscr{H}$ (i.e., $\mathscr{H} = \pm 1$), describes the rotation sense of the in-plane components (m_1 and m_2) of the vortex. Here, $\mathscr{H} = 1$ corresponds to the counterclockwise vortex, while $\mathscr{H} = -1$ corresponds to the clockwise one. Also, $\mathscr{H} = \mathrm{sgn}(\chi)$. More examples that illustrate the difference between these two quantities can be found in Fig. 5.6. For example, in Fig. 5.6a, the $\mathscr{C} = -1$ skyrmion has two forms: either a $\lambda = 1$, $\mathscr{H} = -1$ vortex, or one with the reversed polarity and helicity. Consequently, $\mathscr{C} = \lambda \mathscr{H}$.

The next quantity that is of great importance is the helicity angle χ. Note that we specifically call it the 'helicity angle', instead of simple 'helicity', in order to emphasise its continuous nature. First, helicity is only the sign of the helicity angle, while the value of χ can be continuous. Second, although it has been recognised that Bloch-type skyrmions have $\chi = \pm\frac{\pi}{2}$ [1], while Néel-type skyrmions have $\chi = 0$ or $\chi = \pi$ [2], these are not the only 'quantised' values of χ. In fact, χ can take any arbitrary value from $0°$ to $360°$, though such phases have not been reported yet.

On the other hand, from TDP, it is clear that if a structure is chiral, circular dichroism may be more suitable to study it, as linear-polarisation will smear out the chirality information. Thus, we will concentrate on CD in this chapter.

Further, it is important to remind the reader of the equivalence of the two ways to construct a $N = 1$ skyrmion vortex. As discussed before, a skyrmion can be regarded as a one-dimensional spin helix encircling around a centre once (see Fig. 4.1a). Also, the $N = 1$ skyrmion has the same polarisation dependence as that of a one-dimensional helix. Therefore, the helicity angle of the skyrmion is naturally encoded in the structural property of the one-dimensional spin spiral. An apparent example was illustrated in Chap. 4 where the Bloch-skyrmion was shown to correspond to

© Springer Nature Switzerland AG 2018
S. Zhang, *Chiral and Topological Nature of Magnetic Skyrmions*,
Springer Theses, https://doi.org/10.1007/978-3-319-98252-6_5

the Bloch-type helix, while the Néel-skyrmion corresponds to the Néel-type helix. Therefore, we will start by looking at the representation of an arbitrary helix. This also allows us to derive the analytical solutions without a mathematical challenge. The Fourier transform of a rigid skyrmion solution is purely a mathematical problem, which we try to avoid to solve in this work. Instead, numerical calculations were always performed. Once we can solve the helix structure with REXS, the skyrmion helicity angle is automatically solved.

5.1   Representation of the Arbitrary Spin Spiral

Chiral structures are common, yet some of the most fascinating structures in nature [3]. In condensed matter physics, the emergence of spin chirality is also a common phenomenon, playing an important role governing the magnetic and electronic behaviour of these system, and further allowing for remarkable applications [1, 4, 5]. One prominent example of chiral spin order is the one-dimensional spin helix, in which the magnetic moment modulates along a certain direction, forming a long-wave-length density wave that is usually incommensurate with the crystal lattice. This simple structure can be described by the one-dimensional spin harmonic model, and is found in many different materials, including ferromagnets [6], helimagnets [1], ferroelectric materials [5, 7], geometrical frustrated systems [8], and so on. Such modulated magnetic structures provide an excellent playground for studying the rich underlying physics that governs magnetic order. Furthermore, spintronics applications are proposed based on the materials that carry spiral orders [9]. Therefore, the accurate measurement of the detailed chiral structure of a material is of great importance.

Nevertheless, the unambiguous determination of the spin modulation is a non-trivial issue due to the limitations of the state-of-art magnetic structural characterisation techniques. The periodic helix structure can be described by the modulation wavevector $\mathbf{q}_h$ and the motif. Despite of the chirality degree of freedom, the details of the motif can come in many variations, such the proper-screw-type, cycloidal-type, or conical-type helices [5, 7]. This large motif variety represents a challenge for many experimental techniques. For example, magnetic microscopy methods are usually limited by the spatial resolution, and the inaccessibility to all three components of the magnetisation vector. On the other hand, neutron or x-ray diffraction experiments only accurately measure the wavevector, while the determination of the motif structure is heavily dependent on the diffraction intensity refinement processing, assisted by theoretical calculations [8, 10–21], leaving no unique answer.

Here, we will demonstrate that such modulated spin spiral structures give rise to unique circular dichroism properties in REXS. Due to the sensitivity of the circularly polarised soft x-rays to the chiral magnetisation ordering, the motif information that is encoded in the phase of the Fourier components in the scattering structure factor can be fully retrieved by carrying out geometry-dependent studies. In particular geometries, the circular dichroism goes completely extinct. Such extinction

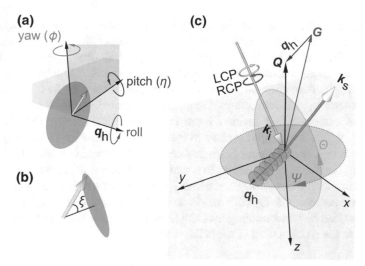

Fig. 5.1 General representation of an arbitrarily modulated spin structure and the measurement geometry that determines its structure. **a** An arbitrarily modulated spin spiral can be represented by a propagation vector $\mathbf{q}_h$, and a common plane within which the spins rotate. We define a zero-position common rotation plane (CRP) as sketched in the figure, for which the plane normal is along $\mathbf{q}_h$. The plane can take pitch, yaw and roll rotation degrees of freedom, giving rise to different types of spin spirals. Therefore, using the combination of two angles η and ϕ, an arbitrary CRP can be represented. In other words, if the set of angles (η, ϕ) can be measured, the exact type of the spin helix can be determined. **b** Once the CRP is determined, a conical angle ξ is used to express the tilt of the spins towards the CRP. However, ξ does not change the type of the spin spiral. **c** To measure the exact CRP, the resonant magnetic diffraction condition at $\mathbf{q}_h$ has to be satisfied, while circularly polarised x-rays are used, in order to measure the circular dichroism. The long-wavelength approximation specifies that $\mathbf{q}_h \ll \mathbf{G}$, in which $\mathbf{G}$ is a structural peak. Two geometrical scans can be carried out, which essentially drive $\mathbf{q}_h$ to rotate along the azimuthal axis Ψ, and the polar axis Θ

condition have a one-to-one correspondence to the motif structure. This forms the basic principle for the unambiguous determination of the modulated spin structure.

For an arbitrary one-dimensional spin harmonic structure, the motif can be represented by a propagation vector $\mathbf{q}_h$ and a common rotation plane (CRP), within which the spins rotate while propagate, as shown in Fig. 5.1a. We define a 'zero-position' of the CRP, also called *base helix*, such that it is normal to $\mathbf{q}_h$. As sketched in Fig. 5.1a, the CRP can take all three rotational degrees of freedom that allows for a rotation along pitch (η), roll, and yaw (ϕ) axes. Therefore, using the polar angle (ϕ, η), the CRP can be uniquely represented. Next, as shown in Fig. 5.1b, the conical angle ξ is assigned, with $0° \leq \xi \leq 90°$, describing the titling angle of the spins with regards to the CRP. Thus, using a set of quantities, (ϕ, η, ξ), a specific type of spin helix can be defined, and all types of such a group of structures are uniquely specified within this representation. In other words, if these three quantities can be measured experimentally, the detailed spin motif can be unambiguously determined.

The experimental REXS geometry necessary to measure those quantities is shown in Fig. 5.1c. The resonant magnetic diffraction condition for the incommensurate wavevector $\mathbf{q}_h$ that can be found as the satellite in reciprocal space is always meet. This is usually achieved by satisfying the diffraction condition for the scattering wavevector $\mathbf{Q}$, where $\mathbf{Q} = \mathbf{G} + \mathbf{q}_h$, and $\mathbf{G}$ is a crystalline Bragg peak. We assume the long-wavelength approximation again, such that $q_h \ll G$. Circular dichroism in this work is again defined as the diffraction intensity difference for the same geometry and wavevector $\mathbf{q}_h$, between the two scattering events using left-circular polarised and right-circular polarised incident light. Two geometrical scans of the CD can be performed, which are complementary to each other, in order to retrieve the values of ϕ and η: (1) Azimuthal circular dichroism (ACD), for which $\mathbf{q}_h$ is placed within the x-y sample plane, and is driven to rotate azimuthally at Ψ from $0°$ to $360°$. (2) Polar circular dichroism (PCD), for which $\mathbf{q}_h$ is placed within the xz-plane, and is driven to rotate about the Θ axis from $0°$ to $360°$. The two scans that orient $\mathbf{q}_h$ can be implemented by either rotating the sample, or rotating the external magnetic field that induces the reorientation of the helix, if feasible.

5.2 Writing Down the Circular Dichroism

Next, we will work out the CD signal as a function of Ψ (or Θ) for an arbitrary spin helix. Equation (4.7) suggest that $I_{CD}(\mathbf{Q})$ takes the form of:

$$I_{CD}(\mathbf{Q}) = 2|F_1|^2 \text{Im}[(\mathbf{k}_s \cdot \mathbf{M}^*(\mathbf{Q}))(\mathbf{k}_s \times \mathbf{k}_i) \cdot \mathbf{M}(\mathbf{Q})] . \tag{5.1}$$

Therefore, the imaginary part of the Fourier components, which usually encode the motif's helicity information and vanish for a diffraction experiment, are visible for CD.

Let us first write up the *base helix* of the motif. Here, it is a proper-screw helix, while $\mathbf{q}_h$ propagates along $\mathbf{x}$ direction, expressed by Eq. (1.15):

$$\begin{aligned}
m_1 &= M_S \cos\xi , \\
m_2 &= M_S \sin\xi \cos(\mathbf{q}_h \cdot \mathbf{r}) , \\
m_3 &= M_S \sin\xi \sin(\mathbf{q}_h \cdot \mathbf{r}) .
\end{aligned} \tag{5.2}$$

Note that there should be $\cos(\mathbf{q}_h \cdot \mathbf{r} + \kappa)$ or $\sin(\mathbf{q}_h \cdot \mathbf{r} + \kappa)$ terms that describe the real-space phase shift κ of the spins within the motif. This will also appear in the expressions of the Fourier transforms. However, this phase information will vanish when calculating the circular dichroism intensity by plugging into Eq. 5.1. This is because one squares the scattering amplitude (due to the famous 'phase problem'), leading to no difference between different κ's. Therefore $\kappa = 0$ is chosen to simplify the derivation.

Next, the arbitrary rotation angles η and ϕ based on this helix will be applied. This corresponds to the combined rotation operations $\mathscr{R}_\phi \mathscr{R}_\eta \mathbf{m}$, where

$$\mathscr{R}_\phi = \begin{pmatrix} \cos\phi & -\sin\phi & 0 \\ \sin\phi & \cos\phi & 0 \\ 0 & 0 & 1 \end{pmatrix}, \text{and } \mathscr{R}_\eta = \begin{pmatrix} \cos\eta & 0 & -\sin\eta \\ 0 & 1 & 0 \\ \sin\eta & 0 & \cos\eta \end{pmatrix}. \text{After this, the helix becomes:}$$

$$m_1 = -M_S \sin\xi \cos\phi \sin\eta \sin(\mathbf{q}_h \cdot \mathbf{r}) - M_S \sin\xi \sin\phi \cos(\mathbf{q}_h \cdot \mathbf{r}) + C_1 ,$$

$$m_2 = -M_S \sin\xi \sin\phi \sin\eta \sin(\mathbf{q}_h \cdot \mathbf{r}) + M_S \sin\xi \cos\phi \cos(\mathbf{q}_h \cdot \mathbf{r}) + C_2 , \quad (5.3)$$

$$m_3 = M_S \sin\xi \cos\eta \sin(\mathbf{q}_h \cdot \mathbf{r}) + C_3 ,$$

where C_1, C_2 and C_3 are ξ-functions which are $\mathbf{r}$-independent. As we are going to perform a Fourier transform later, these terms will become delta functions, which are only non-zero at the origin of the reciprocal space $\mathbf{q} = 0$. Thus, their actual forms are not important at this point.

For ACD, $\mathbf{q}_h$ is required to propagate within the q_x–q_y plane. We define $\Psi = 0°$ as $\mathbf{q}_h$ propagating along the positive q_x direction. In real space, this corresponds to the helix pitch propagating along x. Therefore, in our scattering coordinate system, a rotation of the helix by Ψ will give rise to the magnetisation profile of $\mathscr{R}_\Psi \mathbf{m}$ (see Fig. 4.1b), where $\mathscr{R}_\Psi = \begin{pmatrix} \cos\Psi & -\sin\Psi & 0 \\ \sin\Psi & \cos\Psi & 0 \\ 0 & 0 & 1 \end{pmatrix}$. Imposing this operation again on Eq. (5.3), and performing the Fourier transform on $\mathbf{q} = \mathbf{q}_h$ will lead to:

$$M_1(\mathbf{q}_h) = -\pi M_S \sin\xi \sin(\Psi + \phi) + i\pi M_S \sin\xi \sin\eta \cos(\Psi + \phi) ,$$

$$M_2(\mathbf{q}_h) = \pi M_S \sin\xi \cos(\Psi + \phi) + i\pi M_S \sin\xi \sin\eta \sin(\Psi + \phi) , \quad (5.4)$$

$$M_3(\mathbf{q}_h) = -i\pi M_S \sin\xi \cos\eta .$$

Plugging this into Eq. (5.1) will eventually yield the expression for ACD:

$$I_{\text{ACD}}(\Psi) = \mathscr{C} Y \sin^2\xi \left[\cos\alpha \sin\eta + \sin\alpha \cos\eta \cos(\Psi + \phi)\right] ,$$
$$\propto \pm \left[\cos\alpha \sin\eta + \sin\alpha \cos\eta \cos(\Psi + \phi)\right] , \quad (5.5)$$

where $Y = 4|F_1|^2 \pi^2 k^2 M_S^2 \cos\alpha \sin\alpha$, and α is the angle that satisfies the diffraction condition. $\mathscr{C} = \pm 1$ that describes the chirality of the base helix. For our subsequent calculation, we take the $\mathscr{C} = -1$ spiral as the base helix. Equation (5.5) specifies what is to be expected of the CD when rotating the helix azimuthally, and it is proportional to the combination of a few helix quantities. From a first glance, one should expect that the ACD profile modulates one period as a sinusoidal curve, when rotating Ψ by 2π. The yaw angle ϕ will effectively shift the profile along Ψ, and the pitch angle η will introduce the asymmetry of the sinusoidal curve. Therefore, by fitting the measured ACD profile to Eq. (5.5), the CRP can be unambiguously determined.

To further interpret the ACD, let us first assume $\eta = 0°$, $\phi = 0°$; meaning that the structure is the base helix, as shown in Fig. 5.2a. In this case, Eq. (5.5) reduces to Eq. (4.13) for $N = 1$, identical to the TDP result. The numerical calculation results for the ACD is plotted as the blue curve in Fig. 5.2b. This is simply a cosine curve, being consistent with the results of Eq. (4.13), as shown in Fig. 4.2e. There are two characteristic points, $\Psi = 90°$ and $270°$ at which the ACD becomes completely extinct. If the chirality of the helix is reversed, the sign of the profile is reversed (see red ACD curve).

Let us define the *first extinction condition*, $\Psi_{ext}^{(1)}$, such that $\partial I_{ACD}/\partial \Psi|_{\Psi=\Psi_{ext}^{(1)}} > 0$, meaning that it satisfies two conditions: (1) $I_{ACD}(\Psi_{ext}^{(1)}) = 0$; (2) I_{ACD} should always evolve from negative to positive when crossing over $\Psi_{ext}^{(1)}$. Therefore, in this case, $\Psi_{ext}^{(1)} = 90°$ for the blue curve. It is then obvious that the structure parameter ϕ for the helix directly relates to the first extinction condition by $\phi = \pi/2 - \Psi_{ext}^{(1)} = 0°$, which is consistent with the numerical results. On the other hand, for the red curve, $\Psi_{ext}^{(1)} = 270°$, giving rise to $\phi = -180°$. This suggests that the real-space base helix in Fig. 5.2a should rotate the yaw angle to $-180°$ azimuthally, which is essentially equivalent to the chirality reversal. Therefore, the first extinction condition can fully resolve the information of the yaw angle.

Let us now switch on ϕ, while keeping $\eta = 0$. In this case,

$$I_{ACD} \propto -\cos(\Psi + \phi) , \quad \text{with the first extinction condition:}$$
$$\phi = \frac{\pi}{2} - \Psi_{ext}^{(1)} . \tag{5.6}$$

Note that one can always compress ϕ such that $\phi \in [0, 2\pi)$, as this 2π range contains the complete group elements for the ϕ degree of freedom. On the other hand, $\Psi_{ext}^{(1)} \in [0, 2\pi)$. Therefore, by the linear relation from Eq. (5.6), the ACD has a one-to-one correspondence between the yaw angle and the measured signal. This clearly specifies an unambiguous way that can determine the CRP of the helix. Nevertheless, a negative value of ϕ also reflects the correct description of the helix, thus we will keep the computed ϕ for simplicity. The example is shown in Fig. 5.2d and e, in which the base helix has a yaw canting. The numerical calculation (Fig. 5.2e) shows that $\Psi_{ext}^{(1)} = 60°$, which further predicts that $\phi = 30°$. This is in excellent agreement with Fig. 5.2d. The other example with $\phi = 90°$, $\Psi_{ext}^{(1)} = 0°$ is also shown in Fig. 5.2g and h.

We will now increase the complexity by switching on η. First, as shown in Eq. (5.5) one can ignore the contribution from α, as this is rather a constant which only relates to the long-range-ordered properties. Second, the direct consequence by adding the η degree of freedom is the scaling and shifting effect on the cosine curve, while the periodicity and phase of this cosine profile is undisturbed as if $\eta = 0$. In other words, this does not affect determining ϕ by the first extinction condition using Eq. (5.6) at all: the only extra effort one has to perform is to shift this sinusoidal curve symmetrically, in order to locate $\Psi_{ext}^{(1)}$.

Let us define the *second extinction condition*, $\Psi_{ext}^{(2)}$, at which $I_{ACD} = 0$ by taking η into account. This results in:

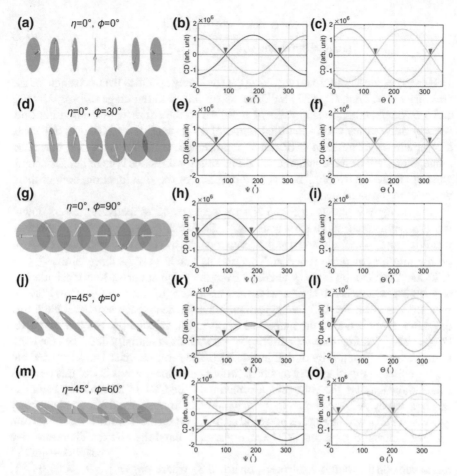

Fig. 5.2 Demonstration of the one-to-one correspondence between the CRP and the dichroism extinction condition. **a** The spin structure with $\eta = 0°$, $\phi = 0°$, otherwise called proper-screw spin helix. **b** The corresponding azimuthal dichroism for **a** is calculated (solid blue curve). The first dichroism extinction condition is labelled by the green triangles. This Ψ value directly correlates to the yaw angle ϕ of the CRP. The asymmetry of the CD profile indicates the value of η. These are evidenced in the subsequent figures. The dashed red curve shows the azimuthal dichroism for the opposite chirality of that in **a**. **c** The corresponding polar circular dichroism for **a** is calculated (solid yellow curve). The dichroism extinction condition is labelled by the green triangles. This Θ value directly correlates to the pitch angle η of the CRP. While the polar circular dichroism is always symmetric, the amplitude of the CD indicates the value of ψ. These are evidenced in the rest of the figures. The dashed magenta curve is for the opposite chirality of that in **a**. **d–o**, The azimuthal dichroism and the polar dichroism calculated for different typical spin spirals.

$$\cos\alpha \, \sin\eta = -\sin\alpha \, \cos\eta \, \cos(\Psi_{ext}^{(2)} + \phi) \,, \quad \text{therefore,}$$
$$\tan\eta = \tan\alpha \, \sin(\Psi_{ext}^{(2)} - \Psi_{ext}^{(1)}) \,. \tag{5.7}$$

Here, $\eta = \tan^{-1}[\tan\alpha \, \sin(\Psi_{ext}^{(2)} - \Psi_{ext}^{(1)})]$ uniquely specifies the pitch angle of the arbitrary helix. First, Eq. (5.7) indicates that η falls into the range of $[-\pi/2, \pi/2]$, and the special conditions $\eta = \pm\pi/2$ will lead to the entire I_{ACD} being a flat line. Also, η only varies within the total range of π, instead of 2π. Only in this case, a complete set of possible CRP can be represented. If both η and ϕ fall into a 2π range, there will be a redundant double counting for the same type of helix. Second, although there are two $\Psi_{ext}^{(2)}$ points, as a property of the sine function, both of them will lead to the same η.

The example is shown in Fig. 5.2j, in which $\eta = 45°$ effectively 'lowers down' the blue sinusoidal curve shown in Fig. 5.2k. We first locate the first extinction condition, $\Psi_{ext}^{(1)} = 90°$, labelled as the green triangle. It is obtained by biasing the I_{ACD} profile into a symmetric function. This reveals that $\phi = 0°$. Subsequently, $\Psi_{ext}^{(2)} \approx 152°$ and $206°$ can be directly obtained from the original curve. In our calculation, $\alpha \approx 48.2°$, which is consistent with the parameters of Cu_2OSeO_3. This leads to $\eta \approx 45°$ based on Eq. (5.7), in agreement with the real-space structure. Note that the minor error is due to the numerical errors in the calculation. For the red curve, $\Psi_{ext}^{(1)} = 270°$, leading to $\phi = -180°$, $\eta \approx -45°$. This essentially describes the helix with the opposite chirality of that in Fig. 5.2j. The same exercise can be carried out for the spiral in Fig. 5.2m and n, which further confirms the validity of this method.

As shown above, the complete information of the CRP of an arbitrary helix can be retrieved by ACD. Nevertheless, we would like to show another complementary geometry, PCD, which is not as powerful as ACD, yet further confirms the result from ACD. As introduced before, PCD actually scans around the Θ-axis. Therefore, one has to perform the rotation matrix $\mathscr{R}_\Theta$ on Eq. (5.3), where $\mathscr{R}_\Theta = \begin{pmatrix} \cos\Theta & 0 & -\sin\Theta \\ 0 & 1 & 0 \\ \sin\Theta & 0 & \cos\Theta \end{pmatrix}$. Subsequently, the Fourier transform on $\mathbf{q} = \mathbf{q}_h$, and the calculation based on Eq. (5.1) results in the expression of the PCD:

$$I_{PCD}(\Theta) = \mathscr{C} Y \sin^2\xi \, [\cos\phi \, \cos\eta \, \sin(\Theta + \alpha) - \sin\eta \, \cos(\Theta + \alpha)] \,,$$
$$\propto \pm[\cos\phi \, \cos\eta \, \sin(\Theta + \alpha) - \sin\eta \, \cos(\Theta + \alpha)] \,. \tag{5.8}$$

Again, $\mathscr{C} = -1$ corresponds to our numerical calculations. Therefore, the extinction condition reads:

$$\tan\eta = \cos\phi \, \tan(\Theta_{ext} + \alpha) \,. \tag{5.9}$$

It is not a straightforward geometry that allows one to retrieve η and ϕ by only observing the extinction condition. Instead, a data fitting process has to be performed. Nevertheless, PCD becomes very useful for certain types of helices that are not too complex. For example, if $\phi = 0°$, Eq. (5.8) reduces to $\eta = \Theta_{exc} + \alpha$, which directly gives rise to the pitch angle. If $\eta = 0°$, the PCD profile will not undergo any phase

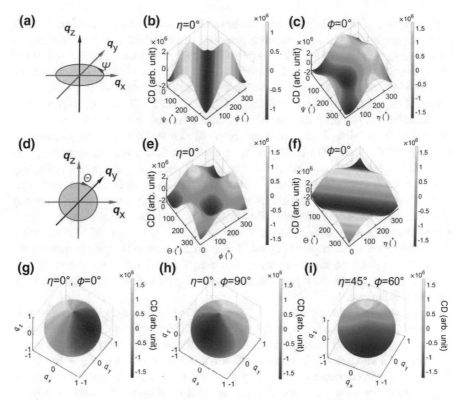

Fig. 5.3 The general features of the dichroism extinction effect. **a** Azimuthal dichroism geometry, in which the Ψ-scan is performed by rotating $\mathbf{q}_h$ by 360° within the q_x–q_y plane. **b** For $\eta = 0°$, each distinct yaw angle ϕ has a unique azimuthal dichroism profile. This allows for the unambiguous determination of ϕ. This is also valid for arbitrary values of η. **c** For $\phi = 0°$ instance, each pitch angle η also has a one-to-one correspondence to the azimuthal dichroism. **d** The measurement geometry for polar dichroism. **e–f** The polar dichroism for different ϕ and η angles, for $\eta = 0°$ and $\phi = 0°$ cases, respectively. **g–i** The entire reciprocal space dichroism for three different spin spirals. It is shown as the dichroism map on a sphere. This map has a one-to-one correspondence to a specific type of spin spiral

shift, and only the amplitude will scale. These results are shown in Fig. 5.2c, f, i, l and o.

In real systems, there are not many types of spin spirals that take very complicated CRPs. For most of the cases, only one degree of freedom (either η or ϕ) may play a role. In this case, PCD becomes very useful.

A summary of the two dichroism geometries for some special helices is shown in Fig. 5.3a–f, in which the one-to-one correspondence between the structural parameter (ϕ or η) and the scan parameter (Ψ or Θ) is established. This also reflects the key idea of carrying out the polarisation-dependent study: one has to let the x-rays 'see' the same magnetisation motif from different angles. The more angles one can measure, the more information that one can retrieve. This also explains why the two geometrical scans give rise to quite different relationships. On the other hand, this points towards to a more general fact: the propagation vector $\mathbf{q}_h$ can be oriented along any direction on a sphere that is centred at the Bragg peak $\mathbf{G}$, forming an entire CD sphere: ACD and PCD are only two circles on it. This is shown in Fig. 5.3g–i for some typical helices. As can be found in these spheres, each arbitrary helix corresponds to a unique CD spherical map, while ACD is the equator on this map, and PCD is a special meridian.

Nevertheless, mapping the entire CD sphere is indeed redundant from an experimental point of view. First, not every material can support this full geometrical mapping. If the spiral can be driven by a magnetic field (like the conical spirals in $P2_13$ helimagnets), as well as that the external field can be applied along any direction, such scan may be possible. However, this is not quite practical for a real instrument, and, in fact, such diffractometer does not exist yet. Moreover, not many materials have a field-driven spin helix. Second, one does not need complete coverage of the CD sphere at all in order to retrieve η and ϕ. In fact, ACD is sufficient for this purpose. Most importantly, ACD can be implemented by purely rotating the azimuthal angle of the sample in the goniometer. This is feasible for almost all systems. Therefore, ACD and its associated structural determination principle is a more universal technique.

Next, we will address the role of ξ played in the CD process. As given in Eqs. (5.5) and (5.8), ξ only scales the amplitude of the CD for any geometrical conditions. It also suggests that it is only the CRP that defines a specific type of spin spiral. The conical spiral actually belongs to the same category as that of a $\xi = 90°$ helix. Therefore, for an unknown modulated spin structure, one can first determine the CRP, regardless of the value of ξ. Afterwards, one will need to check if this modulation is helical or conical. One very quick way to check this information is to look at the $\mathbf{q} = 0$ reciprocal space point. As shown in Eq. (5.3), the constants C_1, C_2, C_3 will be zero if $\xi = 90°$. This will lead to zero magnetic diffraction intensity at $\mathbf{q} = 0$. In other words, if this helix has a conical angle, one will see a non-zero intensity at the magnetic reciprocal space origin. This reciprocal space point overlaps with the structural peaks. Preferentially, a crystallographically forbidden peak would be ideal, as it separates the magnetic origin from the charge peaks. However, an allowed peak would also be possible if a temperature- or field-dependent measurement can be performed.

The influence of ξ is demonstrated in Fig. 5.4, in which the ACD and PCD profiles for different conical angles are plotted together for comparison. Indeed, ξ does not affect the fundamental properties specified by extinction conditions at all, but only scale the amplitude. The larger the conical angle is, the smaller the CD signal that can be measured. When $\xi = 0°$, the CD smears out, as a ferromagnetic-like order is reached, suppressing diffraction at $\mathbf{q} = \mathbf{q}_h$ (even if there is diffraction, the circular dichroism should be zero as the structure is not chiral).

Nevertheless, for the $P2_13$ helimagnets, or other materials that carry both helical and conical orders with same modulation pitch and CRP, the measurement of ξ is possible. For the helical phase, $\xi = 90°$, giving rise to a reference CD signal I_0. Therefore, at the same geometrical angle Ψ or Θ, an applied field will drive the helix into the conical order, with the same CRP. The measured CD signal I will then relate

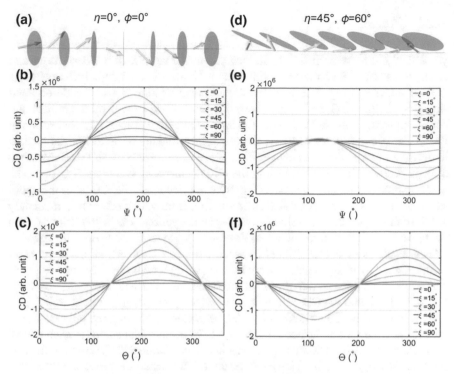

Fig. 5.4 Conical spirals and the related dichroism extinction conditions. **a** A $\eta = 0°$, $\phi = 0°$ conical spiral, otherwise called longitudinal conical structure, is essentially the same type as a proper-screw helix, however, with a $0° < \xi < 90°$ tilt angle. **b–c** For any allowed ξ angle, both azimuthal and polar dichroism keep the extinction condition unchanged with varying ξ. The amplitude of the CD indicates the value of ξ. These features are universal for all conical structures, as illustrated by another spiral example, shown in **d–f**

to the structure via:

$$\frac{I}{I_0} = \sin^2 \xi \ , \quad \text{therefore,}$$

$$\xi = \sin^{-1} \left(\sqrt{\frac{I}{I_0}} \right) \ . \tag{5.10}$$

In brief, by carrying out geometrical scans, the magnetic scattering structure factor can be examined in great depth. This is due to the 'vector sensitivity' of REXS. As a result, the information about the motif structure, i.e., the CRP of the harmonic modulation can be unambiguously determined. Depending on the material's properties, some of them may allow the further measurement of ξ. Once the set of angles (ϕ, η, ξ) is determined, the motif structure is fully resolved.

5.2.1 Results

We will now show experimental data that proves the validity of our derivations and calculations. The REXS experimental setup is shown in Fig. 5.1c, which resembles the one shown in Chap. 4. The material under investigation is again [001]-oriented single crystalline CuO_2SeO_3. However, this time we will focus on the helical and conical phases. In the helical state, the spin modulation is naturally locked along $\langle 001 \rangle$. Therefore, in our geometry, there will be two pairs of helical peaks located in the q_x–q_y plane. This is shown in Fig. 5.5a, measured at 20 K and 0 mT. We then directly perform the dichroism using the CCD camera, as shown in Fig. 5.5b. The first impression from this CD map is that for one Friedel pair, the CD takes the opposite sign. This is a very general feature for CD, from which one can immediately tell the chirality of the spin helix. For example, the yellow peak with positive CD in the upper-right corner corresponds to $\Psi \approx 45°$, while its Friedel pair with negative CD corresponds to $\Psi \approx 225°$. Let us assume this is a standard proper-screw helix (as confirmed by many other reports); by comparing it with Fig. 5.2b, one can conclude that it fits into the red curve, which takes the opposite chirality as shown in Fig. 5.2a. Note that the CD pattern here looks very similar to the polarised SANS pattern reported in [22–25]. Also, in a SANS setup, by rotating the polarisation of the incident neutrons at the helical state of MnSi, a sinusoidal curve for each helical peak around $(0, 0, 0)$ reciprocal space origin can be obtained [26], revealing the chirality of the magnetic helix. However, the underlying scattering process from REXS is fundamentally different from polarised SANS, as can be evidenced in Eq. (2.18). As discussed before, the vector sensitivity of REXS will reveal deeper structural information about the spin motif when carrying out the geometrical scans.

In order to perform the geometrical scans, we can only rotate the azimuthal angle Ψ. This is because, (1), $\mathbf{q}_h$ cannot be driven by the external field, and, (2), rotating about Θ is not possible due to the lack of a rotation centre. Note that if one wants to rotate Θ, the only possible axis is Ω. However, the rotation centre of Ω is at $(0,0,0)$, not $(0,0,1)$. The ACD is then measured and plotted in Fig. 5.5c. Now we can apply the principle we have derived to retrieve its CRP. First, the ACD resembles a simple sinusoidal curve, which is symmetric. The maximum CD is nearly the same as the minimum. This suggests that $\eta = 0°$. Next, the first extinction condition reads $\Psi_{ext}^{(1)} \approx 90°$, leading to $\phi = 0°$. Therefore, by only looking at the data by eye, one can conclude that the helix has a CRP that is the same as that of a proper-screw spiral. Next, the magnetic diffraction (using linear-polarised incident light) at the $(0,0,1)$ peak has to be checked. No temperature dependence of the $(0,0,1)$ diffraction intensity was found when scanning across T_c, meaning that $\xi = 90°$. Therefore, the helical phase structure is unambiguously determined.

A similar measurement on the conical phase was performed, as summarised in Fig. 5.5d–f. First, as expected, the conical structure has the same chirality as that of the helical state. Second, it has almost the same ACD profile as that of the helical order. This means that the CRP of the conical spiral also takes $\eta = 0°$, $\phi = 0°$. As the conical modulation can be effectively driven by the magnetic field, PCD can be

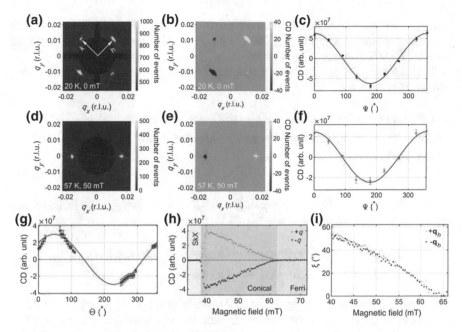

Fig. 5.5 Experimental results from Cu_2OSeO_3 demonstrate the relationship between the spin motifs of the modulated lattice and their dichroism extinction condition. **a** Reciprocal space map (hk-plane at $l = 0$) of the helical state which are at 20 K, 0 mT, measured by resonant x-ray diffraction. Two pairs of the helical peaks are found, due to the lateral domains. **b** In the same state, circular dichroism is performed. The dichroism is obvious for each Friedel pair. It immediately reveals the chirality of the spin helix, which takes the opposite handedness as shown in Fig. 5.2a. **c** Azimuthal dichroism measured for the same state as in **a–b**. The solid red curve is a fitted line using the dichroism extinction condition that takes $\eta = 0°, \phi = 0°$. **d–e** Reciprocal space map (hk-plane at $l = 0$) and the dichroism map for the conical state. The magnetic field is applied along x in real-space. This corresponds to $\Theta = 0°$. The same spin spiral chirality as the helical state can be observed. **f** Azimuthal dichroism of the conical state. As in **c** $\eta = 0°, \phi = 0°$ is used to fit the data, which shows excellent agreement. However, the CD amplitude drops compared with **c**, indicating a tilted conical spiral structure. **g** Polar dichroism result in 50 mT external field. The solid magenta curve is fitted using dichroism extinction condition that takes $\eta = 0°$. **h** Conical CD for the Friedel pair as a function of the external field for $\gamma = 90°$. The evolution of the CD directly correlates with the value of ξ at each magnetic field. Each magnetic phase is labelled, including the skyrmion lattice phase (SkX), conical and ferrimagnetic phase. Note that the blue region is the coexistence region of the SkX and the conical state, which is the phase transition region. **i** Based on **h**, the exact value of ξ can be obtained, according to Eq. (5.10).

performed by rotating the magnetic field with the same amplitude around Θ. The result is shown in Fig. 5.5g, which can be fitted by Eq. (5.8), with $\eta = 0°$, and $\phi = 0°$, as shown as the magenta curve. It is also consistent with the numerical calculation shown in Fig. 5.2c.

If a CCD camera is used for carrying out the geometrical CD scans, actually there is a simpler way. As shown in Fig. 5.5e, when positioning the conical peaks at $\Psi = 0°$, a two-fold magnetic peak can be found within the q_x–q_y plane, corresponding to $+\mathbf{q}_h$

and $-\mathbf{q}_h$. We would like to emphasise that for such Friedel pair, if $+\mathbf{q}_h$ has its ACD position of Ψ; in order to find $I_{\mathrm{ACD}}(\Psi + \pi)$, one actually does not need to rotate the Ψ-axis by 180°. Instead, $I_{\mathrm{ACD}}(\Psi + \pi) = I_{\mathrm{ACD}}(-\mathbf{q}_h)$.

This can be easily proved by looking at Eq. (5.4). By rotating Ψ by π, the Fourier components become:

$$M_1(\Psi + \pi) = \pi M_S \sin\xi \sin(\Psi + \phi) - i\pi M_S \sin\xi \sin\eta \cos(\Psi + \phi) = -M_1(\Psi)$$
$$M_2(\Psi + \pi) = -\pi M_S \sin\xi \cos(\Psi + \phi) - i\pi M_S \sin\xi \sin\eta \sin(\Psi + \phi) = -M_2(\Psi) \ ,$$
$$M_3(\Psi + \pi) = -i\pi M_S \sin\xi \cos\eta = M_3(\Psi) \ .$$

$$(5.11)$$

Also, the property of the Friedel pair gives: $\mathbf{M}(-\mathbf{q}_h) = \mathbf{M}^*(\mathbf{q}_h)$. Evaluating these expressions and inserting them into Eq. (5.1), there is:

$$I_{\mathrm{ACD}}(\Psi + \pi) = I_{\mathrm{ACD}}(-\mathbf{q}_h) = -I_{\mathrm{ACD}}(\Psi) \ , \quad \text{and similarly,}$$
$$I_{\mathrm{PCD}}(\Theta + \pi) = I_{\mathrm{ACD}}(-\mathbf{q}_h) = -I_{\mathrm{ACD}}(\Theta) \ ,$$

$$(5.12)$$

where $\mathbf{q}_h$ is propagating along Ψ for ACD, while propagating along Θ for PCD. Equation (5.12) is extremely useful for carrying out geometrical scans using a CCD camera, as we now only need to rotate the sample by π while observing the CD for a Friedel pair, in order to gather the full ACD (PCD) profile.

Next, we will measure the exact conical angle ξ for the conical phase. Figure 5.5h shows the CD as a function of magnetic field at a specific ACD position of $\Psi = 0°$. At 57 K, a finite magnetic field will first stabilise the skyrmion lattice phase. When increasing the field, the skyrmion lattice phase will undergo a first-order transition into the conical state. A larger field will finally destroy the conical structure, while dragging all the spins to align parallel. Nevertheless, the exact form of the conical structure has not been resolved yet by other characterisation techniques. Here, based on Eq. (5.10), it is shown that if a reference $\xi = 90°$ helical CD is known, based on the conical CD signal, ξ can be unambiguously determined. Note that diffractions have to be performed in the identical geometries, i.e., for the ACD geometry, Ψ has to be kept same.

As shown in Fig. 5.5h, by keeping $\Psi = 0°$ and $\Psi = 180°$ (for each Friedel pair peak), the conical CD as a function of magnetic field can be measured. First, the evolution of the signal reflects the phase transition process. Starting from the skyrmion lattice phase, there is no conical peak at the $\Psi = 0°$ position for $\gamma = 90°$. At around 38–40 mT, the conical peak CD intensity quickly builds up, which corresponds to a sharp conical-to-skyrmion transition. We recorded the helical CD at the same azimuthal angle to be $\sim 5.9 \times 10^7$ (arb. units), as shown in Fig. 5.5c. Therefore, by applying Eq. (5.10) on the measured conical CD, the exact ξ value as a function of field can be determined, as plotted in Fig. 5.5i.

To this point, the complete set of quantities that defines a spin spiral structure, i.e., (ϕ, η, ξ) for the helical and conical phases of a $P2_13$ helimagnet are determined. The resolved structure is unique, with no ambiguous alternatives. In this respective,

REXS differs from other characterisation techniques. Although both ACD and PCD are demonstrated, it is clear that only the ACD is necessary to be performed in order to determine ϕ and η, the two most important quantities that define the CRP. The subsequent activity is to check if the motif is helical or conical by checking the magnetic diffraction in the $\mathbf{q} = 0$ point. One can implement this technique to study a vast variety of spin modulated materials to this level. Further, measurement of the conical angle using the method shown above is not available for all materials. Nevertheless, by comparing the diffraction intensities at $\mathbf{q} = 0$ and $\mathbf{q} = q_h$, ξ can also be effectively measured, although not shown here.

5.3 Measurement of the Skyrmion Helicity Angle

5.3.1 Characteristic Helix and the Associated Skyrmion

Since we have established a technique that retrieves the 'phase' information of the spin helix, it is then very easy to apply the same method to a skyrmion vortex. As we have proven in Chap. 4, the helix structure and skyrmion lattice structure have the identical ACD. In other words, when x-rays 'see' the structure azimuthally, they essentially 'see' the helix and the skyrmions as the same object. This seemingly surprising fact can be easily explained by looking at Fig. 4.1a. At one azimuthal angle Ψ, the CD only checks the magnetic structure that modulates along a specific direction related to Ψ. The structure factor $\mathscr{F}(\mathbf{q}_h)$ is directly related to the Fourier transform of $\mathbf{M}(\mathbf{q}_h)$. As can be expected, any real-space structure $\mathbf{m}(\mathbf{r})$ that does not contribute to the periodicity along $\mathbf{q}_h$ will not be pronounced at $\mathbf{M}(\mathbf{q}_h)$, though this piece of information will be encoded in $\mathbf{M}(\mathbf{q}_h)$. For our ACD scan, we only concentrate on one $\mathbf{q}_h$, i.e., only this wavevector is rotated and measured. Therefore, for one skyrmion lattice peak (one out of six wavevectors), it will look extremely similar as a helical peak, and it has the same azimuthal dependence of a helix wavevector. This is also consistent with the equivalence between the triple-$\mathbf{q}$ solution and axial-symmetric skyrmion solution that is used to describe a skyrmion lattice. As shown in Chap. 3, using the triple-$\mathbf{q}$ ansatz, a general skyrmion lattice can be written as:

$$m_1 = -\frac{1}{3} M_S \sum_{i=1}^{3} \sin(\Psi_i + \phi) \cos(q_h \cos\Psi_i x + q_h \sin\Psi_i y) \ ,$$

$$m_2 = \frac{1}{3} M_S \sum_{i=1}^{3} \cos(\Psi_i + \phi) \cos(q_h \cos\Psi_i x + q_h \sin\Psi_i y) \ , \tag{5.13}$$

$$m_3 = \frac{1}{3} M_S \sum_{i=1}^{3} \sin(q_h \cos\Psi_i x + q_h \sin\Psi_i y) \ ,$$

Fig. 5.6 Illustration of the relationship between the skyrmion texture and its characteristic helix. **a** A $C = -1$ (left-handed) Bloch-type skyrmion can take two different forms. Both forms have the same characteristic helix. **b** If the characteristic helix takes a yaw angle of 30°, the associated skyrmion will change its texture, manifesting itself as a 'diverged' vortex with its core pointing up; or a 'converged' vortex with its core pointing down. **c** A $\phi = -60°$ characteristic helix and its associated skyrmion. **d** Scenario with the opposite chirality as in **c**

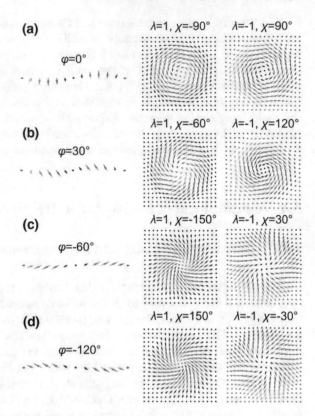

where Ψ_i is the azimuthal rotation of the helix propagation wavevector $\mathbf{q}_h$, with its length q_h. Here, the three helices are of the same type, with identical yaw angle ϕ and zero pitch angle η. As can be seen from Eq. (5.13), the skyrmion lattice with the individual skyrmions, taking an arbitrary helicity angle, can be reproduced by three coherently propagating helices with canted yaw angle. This indicates the fact that the yaw angle of the helix actually shares the same degree of freedom as the helicity angle of a skyrmion.

The other way of looking at this issue is to imagine a skyrmion along a one-dimensional line that passes through its core. Such one-dimensional structure is essentially a spin helix. We then extract this one-dimensional helix as the *characteristic helix*, which encodes all the necessary information we would like to know about the skyrmion vortex. Vice versa, a $\eta = 0$ helix can be 'expanded' into a two-dimensional skyrmion, called *associated skyrmion*. We would like to emphasise that there is a one-to-two correspondence between a characteristic helix and its associated skyrmion vortex.

As shown in Fig. 5.6a, the left-handed Bloch-type skyrmion can come in two appearances with opposite polarity. By extracting the one-dimensional line that passes through the skyrmion core, the characteristic helix is obtained, as shown on the left. It is clear that both two skyrmions have the same type of characteristic helix. This is also true for Fig. 5.6b–d, in which one characteristic helix unambiguously 'stands for' two types of skyrmions with same chirality but opposite polarities.

The polarity reversal, i.e., the skyrmion core reversal, can be achieved by flipping the magnetic field. However, once the core is flipped, the helicity angle has to change accordingly in order to conserve the global chirality. This can be easily understood by looking at the Zeeman energy term $\mathbf{B} \cdot \mathbf{m}$. The odd function requires $\mathbf{m}$ to undergo an inversion symmetry transformation if $\mathbf{B}$ changes its sign. Evaluating Eq. (1.11), the inversion operation is realised by transforming $\lambda \to -\lambda$, and $\chi \to \chi + \pi$, as shown in the figure. Alternatively, or equivalently, in Eq. (5.13), this will be achieved by performing a 'phase shift' such that

$$\cos(q_h\cos\Psi_i x + q_h\sin\Psi_i y) \to \cos(q_h\cos\Psi_i x + q_h\sin\Psi_i y + \pi) \ ,$$
$$\sin(q_h\cos\Psi_i x + q_h\sin\Psi_i y) \to \sin(q_h\cos\Psi_i x + q_h\sin\Psi_i y + \pi) \ .$$

However, ϕ does not change. This means that the magnetic field reversal does not change the fundamental type of the characteristic helix. Therefore, we can conclude that once the characteristic helix is determined, its associated skyrmion type is also determined. For a $\lambda = 1$ skyrmion, its helicity angle relates to the yaw angle of its characteristic helix by:

$$\chi = \phi - \pi/2 \ , \quad \text{polarity reversal } \lambda \to -\lambda \text{ lead to:}$$
$$\chi \to \chi + \pi \ . \tag{5.14}$$

This also reveals the fundamental relationship between the triple-$\mathbf{q}$ model and the axial-symmetric skyrmion model.

Next, let us recall the numerical work performed in Chap. 4, where it was shown that the characteristic helix lattice has the identical ACD as that of the associated skyrmion lattice. Especially, in Eq. (4.16), the special situation with $\mathscr{C} = -1$ and $N = 1$ reduces to Eq. (5.6). It is clear that the 'random' phase Φ_1 that is used to make up the TDP is actually the yaw angle ϕ of its characteristic helix, which in turn relates to the helicity angle χ by Eq. (5.14).

It is important to note here that the exact form of $\mathbf{M}(\mathbf{q}_h)$ must be different for a skyrmion lattice and its characteristic helix lattice, however, the ACD profiles should be the same. This is due to the special sensitivity of the magnetic circular dichroism: it only shows a non-zero signal for any magnetisation modulation that is chiral, and automatically filters other contributions by performing the subtraction operation. This becomes more obvious when evaluating Eq. (5.1), in which only the imaginary part eventually matters. It otherwise reads:

$$I_{CD} \sim a\text{Re}(M_1)\text{Im}(M_2) + b\text{Re}(M_2)\text{Im}(M_1) + c\text{Re}(M_3)\text{Im}(M_2) + d\text{Re}(M_2)\text{Im}(M_3) \ , \tag{5.15}$$

where a, b, c and d are constants. The 'exchange' between the real and imaginary parts of the three Fourier components essentially contains the chirality and helicity information. Furthermore, no other real part terms are present in the equation, which guarantees that the CD is insensitive to other structurally varying details. For example, the detailed $\theta(\rho)$ profile for a skyrmion is the property of the m_3 modulation. One can always use a Fourier series, i.e., a set of sine and cosine functions to fit this profile. Eventually, the chiral information that is determined by the 'exchange' between the Fourier components will be pronounced, while the detailed profile of m_3 will be invisible. This is why we have seen from Chap. 4 that whatever $\theta(\rho)$ function we use, it does not disturb the ACD profile at all.

The concept of a characteristic helix, and the fact that its ACD is identical to that of its associated skyrmion lattice, finally allows us to determine the helicity angle, using the same dichroism extinction condition during the ACD measurement. Furthermore, as the characteristic helix of a skyrmion naturally takes $\eta = 0$, the entire analysis is largely reduced.

5.3.2 Numerical Calculations

We will now show numerical calculations that compute ACD for a skyrmion lattice phase with different helicity angles. This will resemble the identical relationship as plotted in Fig. 5.3b. However, this time, the ACD will be done differently, yet efficiently, compared with how we dealt with the helix lattice. As shown in Eq. (5.13), the parameter of Ψ_i essentially plays the identical role as the azimuthal angle Ψ in the ACD expression. Therefore, Eq. (5.13) suggests that the six magnetic peaks for a skyrmion lattice directly represent six Ψ angles as if an azimuthal scan was performed. Consequently, instead of rotating the sample with varying Ψ, one can measure the CD for six magnetic peaks at the same time, and fit them in Eq. (5.5). By finding the extinction condition, the yaw angle of the characteristic helix can be determined. Furthermore, as the polarity of the magnetic field is a known parameter, λ is known. In this case, χ can be unambiguously measured by referring to Eq. (5.14).

The numerical calculation results are shown in Fig. 5.7. The skyrmion lattices that are generated by both Eqs. (5.13) and (1.11) are used, which show identical ACD. As can be found in the CD patterns in Fig. 5.7, only six CD magnetic peaks were used to replace ACD, which turns out to be sufficient to find the dichroism extinction condition. This method is extremely advantageous from an experimental perspective, as the ACD can be performed using a CCD camera without rotating the azimuthal angle of the sample. In this case, high efficiency and accuracy can be obtained. Rotating the azimuthal angle, on the other hand, will cost more time and introduce certain errors, as the sample can never be perfectly aligned in the goniometer. The simulation results are focusing on the $\lambda = 1$ skyrmion lattice, as shown in the centre of Fig. 5.7. As expected, different helicity angle gives rise to different extinction conditions, described by Eq. (5.6). We then draw an *extinction line* with a direction, which is nothing but connecting the two first extinction conditions as shown in Fig. 5.2. Consequently, such an extinction line has a one-to-one correspondence with χ, forming a 'clock' that uniquely specifies the helicity angle. Therefore, this 'clock' gives us a look-up-table to be compared with the experimental data.

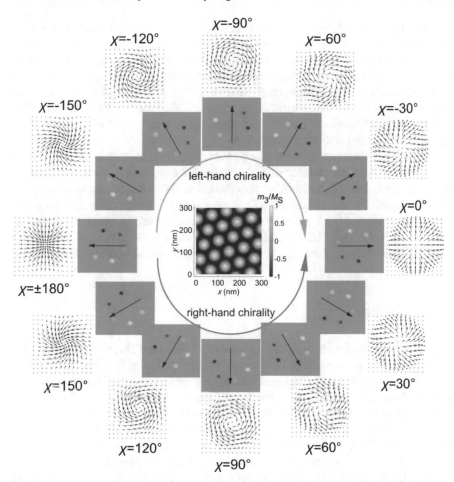

Fig. 5.7 Illustration of $N = 1$ skyrmions with different helicity angles χ, and the associated ACD pattern. The centre figure shows the skyrmion lattice used for the REXS CD calculation, as well as its vortex polarity of $\lambda = 1$. The associated chirality, χ, ACD, and the calculated extinction line are shown. Reprinted from Ref. [27]. Copyright 2018 by American Physical Society

Table 5.1 Possible transformations on the skyrmions and the associated change of the look-up-table

Transformation	Skyrmion texture	$I_{\mathrm{ACD}}(\Psi)$	Characteristic helix (ϕ)
$\lambda \to -\lambda$	Core reversal	$-I_{\mathrm{ACD}}(\Psi)$	$\phi + \pi$
$\chi \to -\chi$	Rotation sense reversal	$I_{\mathrm{ACD}}(\Psi - 2\chi)$	$\pi - \phi$
$\mathscr{C} \to -\mathscr{C}$	Chirality reversal (due to either λ or χ reversal)	$-I_{\mathrm{ACD}}(\Psi)$, or $I_{\mathrm{ACD}}(\Psi - 2\chi)$	$\phi + \pi$, or $\pi - \phi$
$D \to -D$	Global chirality reversal	$I_{\mathrm{ACD}}(\Psi - 2\chi)$	$\pi - \phi$
$\mathbf{B} \to -\mathbf{B}$	Core reversal and $\chi \to \chi + \pi$	No change	No change
$\Psi_i \to \Psi_i + \delta\Psi$	Lattice rotates	No change	No change

From Fig. 5.7 and Table 5.1, it can be seen that the measurement principle is a direct method that is able to identify the exact skyrmion texture with high accuracy. First, the extinction line is independent of the actual skyrmion lattice locking orientation, making the measurement robust against arbitrary sample positioning. Second, in a real experiment, the extra control parameter we can have is the reversal of the magnetic field **B** and the sample's crystalline chirality (denoted as the sign of the DMI). The resulting effects have all been predicted in Table 5.1, which gives excellent guidance for what to expect from the data.

In summary, there are three features for measuring χ for the skyrmion lattice phase. First, this activity is identical to that of measuring the yaw angle of the characteristic helix using the dichroism extinction condition. Second, as the skyrmion lattice phase has six magnetic peaks, one can use a CCD camera to perform circular dichroism all in once. In this case, ACD becomes a reciprocal space pattern, in which the extinction line can be calculated. Third, as the polarity of the magnetic field is known, λ is known. Consequently, the extinction line unambiguously reveals the helicity angle.

5.3.3   Results

Cu_2OSeO_3 carries Bloch-type skyrmions with $\chi = \pm 90°$. Therefore, we expect an ACD pattern with an extinction line that is vertically aligned (see Fig. 5.7). In contrast, the measured pattern in Fig. 5.8a shows a different extinction line. First, the temperature is 57 K, while the magnetic field is 32 mT, with $\gamma = 0°$. The six-fold-symmetric diffraction pattern is assuring that the skyrmion lattice phase (not helical or conical phase) is stabilised. The 20 μm diameter pinhole and the clean pattern confirms that the x-rays probe the single-domain state. The $\gamma = 0°$ angle allows us to confirm that the skyrmions have $\lambda = 1$. By tuning the photon energy to the L_3 edge of Cu, the measured extinction line largely deviates from the vertical direction.

The fitting progress can be found in Fig. 5.8b. First, the six dichroic peaks form a perfect symmetric sinusoidal curve, suggesting that the characteristic helix has $\eta = 0°$. Second, the fitted first extinction condition yields $\Psi_{ext}^{(1)} = \phi = -121°$. This directly gives rise to the extinction line, which relates to $\chi = 149°$. The calculated ACD is shown in Fig. 5.8c, with the same χ. This reveals the fact that the skyrmions are not 'standard' vortices as expected, although their chirality is clearly right-handed. Instead, the vortex structure appears to be more 'converged' as shown in Fig. 5.8d.

This result is in stark contrast to the standard model. It indicates the special features of REXS: its surface sensitivity. Therefore, it may suggest that the skyrmion texture on the surface may be quite different from the bulk. In other words, if the x-rays can penetrate deeper into the bulk of the sample, the helicity angle may recover to $\chi = 90°$. In order to probe deeper into the sample, an above-edge photon energy is used. The sampling depth is shown in Fig. 2.7f. At 931.25 eV, the x-rays 'see' only $\sim$31 nm deep into the sample. While at 932 eV, the penetration depth increases to $\sim$67 nm. The result is shown in Fig. 5.8e–h, in which a less 'converged' skyrmion appears. The fitted helicity angle takes $\chi = 126°$. This suggests a three-

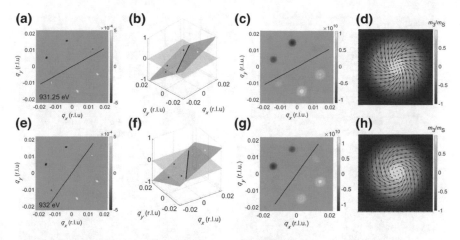

Fig. 5.8 Experimental results that measure χ at different probing depths. **a** ACD pattern on the skyrmion lattice phase of Cu_2OSeO_3. The temperature is 57 K, and the magnetic field is +32 mT, with $\gamma = 0°$. The photon energy is 931.25 eV, which is on the peak of the L_3 absorption edge. The extinction line is fitted and labelled (red line), corresponding to $\chi = 149°$. **b** Three-dimensional view of **a**, in which the six dichroic peaks lie on a common red plane. The $I_{ACD} = 0$ plane is marked as the blue plane. **c** Numerically calculated ACD pattern based on the skyrmion motif shown in **d**, with $\chi = 151°$. **e–h**, same as **a–d**, however with the photon energy above the L_3 edge, with 932 eV. In this case, the x-rays probe much deeper through the sample. The fitted helicity angle is 126°, which is the same used in the calculations in **g–h**.

dimensional skyrmion structure in which the helicity angle changes from the bulk to the surface. In the depth of the sample, the skyrmion is a standard Bloch-type spin swirl. However, when getting closer to the surface, the helicity angle becomes more and more 'converged'. Clearly, 67 nm is not 'deep' enough to reach undistorted skyrmions, thus we still get a helicity angle of 126°.

To gain more insights into the three-dimensional structure, as well as the depth-dependence REXS process, we might need to deal with the attenuation length in a more quantitative way. For a homogeneous sample, the x-ray absorption cross-section σ is calculated using the celebrated multiplet calculations towards Cu L edge [28]. The penetration length Λ, measures the depth scale that drops $1/e$ of the incidence intensity, takes the form of $\Lambda = \sigma^{-1}$. This is plotted in Fig. 5.9a, calculated by Gerrit van der Laan.

For REXS, the scattered intensity is also a function of the incidence angle α_i and outgoing angle α_f of the x-ray beam. Thus, the REXS sampling depth is written as $\Lambda/(\sec\alpha_i + \sec\alpha_f)$. For our experimental setup, the diffraction condition is almost a specular reflection, therefore $\alpha_i = \alpha_f = \alpha$. The REXS sampling depth as a function of photon energy for our sample is plotted in Fig. 5.9b. As a result, for a three-dimensional scattering object, the scattering intensity from a 'layer' z ($z = 0$ corresponds to the surface, and it measures the distance from the surface), taking the form of [29]:

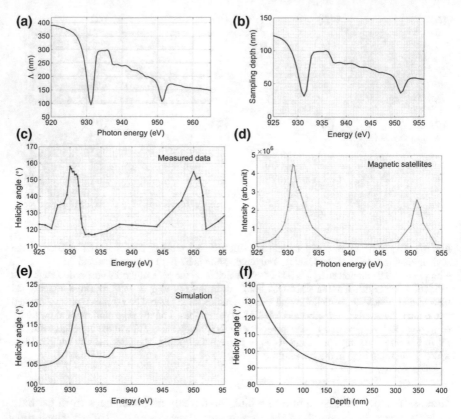

Fig. 5.9 Energy dependent properties. **a** Calculated penetration depth as a function of photon energy. **b** X-ray sampling depth as a function of photon energy for our scattering geometry. **c** Experimentally measured helicity angle as a function of photon energy. **d** Magnetic satellite peak intensity as a function of photon energy. The incident polarisation is π-polarised. **e** Calculated average helicity angle as a function of photon energy, using the magnetisation configuration generated by micromagnetic simulations. The simulated skyrmion helicity angle as a function of z is shown in **f**

$$Y(z) = I_{CD}(z) \frac{1}{\Lambda} \exp\left(-\frac{2z}{\Lambda} \sec\alpha\right) , \qquad (5.16)$$

where I_{CD} is the depth-independent dichroism signal from the depth $z \rightarrow z + \delta z$. Therefore, the total measured CD takes the form of:

$$I_{CD}^{total} = \int_{z=0}^{\infty} Y(z)dz . \qquad (5.17)$$

As can be clearly seen in Eq. (5.17), the measured χ using the dichroism extinction condition is actually an 'averaged' helicity from many individual 'layers'. The deeper we probe, the more $\chi = 90°$ skyrmions will participate in the integration in

Eq. (5.17), which 'neutralise' the averaged helicity angle. Therefore, to better identify the helicity angle evolution as a function of depth, an energy scan has to be
performed. Figure 5.9c shows the measured average helicity angle as a function of
photon energy. Clearly, it resembles the basic profile as the sampling depth (see
Fig. 5.9b). While x-rays probe deeper, the 'converged' helicity angle is more neu-
tralised, giving rise to relatively small χ; and vice versa. This directly suggests that
the skyrmion helicity angles undergoes a smooth, continuous transformation from
bulk to the surface.

Three-dimensional micromagnetic simulation were performed by Jan Müller from
the University of Cologne, from which the helicity angle change as a function of depth
was recovered. As shown in Fig. 5.9f, the skyrmion helicity angle indeed continuously
transforms from standard skyrmions into 'converged' skyrmions, while approaching
from bulk to the surface layer. Using this configuration, the energy-dependence of the
averaged χ can be calculated using Eq. (5.17) together with the dichroism extinction
condition, as plotted in Fig. 5.9e. This is qualitatively consistent with the measured
data to a large degree.

Next, we study the detailed skyrmion texture change by reversing the magnetic
field. This is implemented by rotating the magnetic stage γ by 180° in a well-defined
skyrmion lattice state at 32 mT and 57 K. Figure 5.10a shows the reciprocal space
map of the newly formed skyrmion lattice phase. The extra diffraction spots are
due to another skyrmion domains. This is caused during the field-reversal process.
As discussed in Chap. 3, a magnetic field that deviates from the major crystalline
axes will induce the formation of additional skyrmion domains. As γ is driven by
motors, a continuous rotation with all angles from 0° to 180° will occur during
the field-flipping process. This gives rise to the formation of other skyrmion lattice
domains with different orientations. Nevertheless, as shown before (see Table 5.1),
extra magnetic peaks do not interrupt the finding of the extinction line at all. In
fact, it may even be more clearer to identify the boundary between the positive CD
and negative CD parts around the Ψ circle, from which the extinction line will be
extracted. (Imagine an ACD pattern that consists of a ring, instead of six spots. In
this case, it is very easy to directly draw an extinction line dividing the positive and
negative part. This will be more effective than fitting six points into a sinusoidal
curve.)

Therefore, we can continue carrying out the circular dichroism measurement
without caring too much if the skyrmion lattice state is 'clean' or not. Figure 5.10b
shows the ACD pattern for the identical state to that in Fig. 5.10a, in which the
extinction line is extracted, showing the same line as in the case of $\gamma = 0°$ (see Fig.
5.8a). As expected, the reversal of **B** will not change the extinction condition due to the
odd symmetry of the Zeeman energy term. Given that $\lambda = -1$, it is straightforward
to conclude that $\chi = -34°$, with a real-space texture shown in Fig. 5.10c.

We would like to address the ACD in the multidomain skyrmion state as well.
The result is shown in Fig. 5.10d, in which the multidomain state is created by
field-cooling the system into the skyrmion state with 32 mT, $\gamma = 17°$ tilting field.
In this case, it is much easier to identify the extinction condition as more magnetic
peaks participate in ACD, though with different CD amplitude. Nevertheless, the

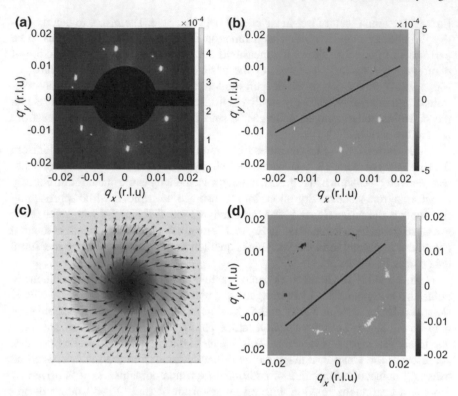

Fig. 5.10 a Reciprocal space map (*hk*-plane at $l = 1$) of the skyrmion lattice phase using π-polarised incident light. The magnetic field is -32 mT for $\gamma = 180°$. The temperature is 57 K. **b** ACD pattern at the same condition of **a**. The photon energy is 931.25 eV. The fitted extinction line corresponds to $\chi = -34°$. This relates to a skyrmion texture as shown in **c**. **d** The ACD pattern for a multidomain skyrmion lattice state. The magnetic field in this case is 32 mT for $\gamma = 17°$. The fitted extinction line is 135°

determined helicity angle is 135°, which is less than that of the $\gamma = 0°$ case. This is because the skyrmion tubes are titled by 17° relative to the sample surface. In this case, with the same scattering geometry, x-rays can probe deeper along the skyrmion tubes. This will effectively reduce the averaged skyrmion helicity angle.

So far we have shown three important facts: (1) On the surface, the skyrmion helicity angle largely deviates from the 'Bloch-type' skyrmion. The detailed spin configuration of the individual skyrmion can be reconstructed. (2) In detail, an averaged skyrmion texture is reconstructed. In fact, the helicity angle evolves continuously from bulk to surface. This depth-dependent profile can be well fitted from the measured energy scan. (3) By reversing the magnetic field, the characteristic helix does not change its structure at all. Consequently, the skyrmion helicity angle will be very different for the two different polarities.

Last, we would like to discuss about the significance of these findings, which have not been reported nor studied so far. First, the continuous nature of the helicity angle indicates that there is an extra degree of freedom for the skyrmions. This quantity can essentially serve as a state variable for memory applications. In this case, the helicity angle can be manipulated by exotic efforts (such as magnetic field), while keeping the skyrmion as an intact, topologically-protected information carrier. Second, its appearance on the surface suggests that the bulk DMI may not be sufficient to account for the formation of the skyrmion structure. In other words, the specific environment at the surface, at which the inversion symmetry is naturally broken, may induce an extra antisymmetric microscopic interaction that explains the extra helicity angle that deviates from $\pm 90°$. Third, from an application point of view, it is only the surface skyrmions (either in thin film or bulk form) that are of great importance for device applications. This provides a roadmap in which surface engineering is the key for further skyrmion research. By adjusting the parameters on the surface level, new complex magnetic structures, as well as the manipulation of the skyrmions, can be achieved.

References

1. N. Nagaosa, Y. Tokura, Nat. Nanotech. **8**, 899 (2013)
2. I. Kezsmarki, S. Bordacs, P. Milde, E. Neuber, L.M. Eng, J.S. White, H.M. Ronnow, C.D. Dewhurst, M. Mochizuki, K. Yanai, H. Nakamura, D. Ehlers, V. Tsurkan, A. Loidl, Nat. Mater. **14**, 1116 (2015)
3. S. Bordács, I. Kézsmárki, D. Szaller, L. Demkó, N. Kida, H. Murakawa, Y. Onose, R. Shimano, T. Rõõm, U. Nagel, S. Miyahara, N. Furukawa, Y. Tokura, Nat. Phys. **8**, 734 (2012)
4. P.M. Chaikin, T.C. Lubensky, *Principles of Condensed Matter Physics* (Cambridge University Press, 1995)
5. Y. Tokura, S. Seki, Adv. Mater. **22**, 1554 (2010)
6. H.A. Dürr, E. Dudzik, S.S. Dhesi, J.B. Goedkoop, G. van der Laan, M. Belakhovsky, C. Mocuta, A. Marty, Y. Samson, Science **284**, 2166 (1999)
7. Y. Tokura, S. Seki, N. Nagaosa, Rep. Prog. Phys. **77**, 076501 (2014)
8. A. Biffin, R.D. Johnson, I. Kimchi, R. Morris, A. Bombardi, J.G. Analytis, A. Vishwanath, R. Coldea, Phys. Rev. Lett. **113**, 197201 (2014)
9. M.N. Wilson, E.A. Karhu, D.P. Lake, A.S. Quigley, A.N. Bogdanov, U.K. Rößler, T.L. Monchesky, Phys. Rev. B **88**, 214420 (2013)
10. F. Fabrizi, H.C. Walker, L. Paolasini, F. de Bergevin, A.T. Boothroyd, D. Prabhakaran, D.F. McMorrow, Phys. Rev. Lett. **102**, 237205 (2009)
11. S.B. Wilkins, T.R. Forrest, T.A.W. Beale, S.R. Bland, H.C. Walker, D. Mannix, F. Yakhou, D. Prabhakaran, A.T. Boothroyd, J.P. Hill, P.D. Hatton, D.F. McMorrow, Phys. Rev. Lett. **103**, 207602 (2009)
12. R.D. Johnson, S. Nair, L.C. Chapon, A. Bombardi, C. Vecchini, D. Prabhakaran, A.T. Boothroyd, P.G. Radaelli, Phys. Rev. Lett. **107**, 137205 (2011)
13. A.J. Hearmon, F. Fabrizi, L.C. Chapon, R.D. Johnson, D. Prabhakaran, S.V. Streltsov, P.J. Brown, P.G. Radaelli, Phys. Rev. Lett. **108**, 237201 (2012)
14. G.E. Johnstone, R.A. Ewings, R.D. Johnson, C. Mazzoli, H.C. Walker, A.T. Boothroyd, Phys. Rev. B **85**, 224403 (2012)
15. R.D. Johnson, P. Barone, A. Bombardi, R.J. Bean, S. Picozzi, P.G. Radaelli, Y.S. Oh, S.-W. Cheong, L.C. Chapon, Phys. Rev. Lett. **110**, 217206 (2013)
16. R.D. Johnson, K. Cao, L.C. Chapon, F. Fabrizi, N. Perks, P. Manuel, J.J. Yang, Y.S. Oh, S.-W. Cheong, P.G. Radaelli, Phys. Rev. Lett. **111**, 017202 (2013)

17. A.J. Hearmon, R.D. Johnson, T.A.W. Beale, S.S. Dhesi, X. Luo, S.-W. Cheong, P. Steadman, P.G. Radaelli, Phys. Rev. B **88**, 174413 (2013)
18. H.C. Walker, F. Fabrizi, L. Paolasini, F. de Bergevin, D. Prabhakaran, A.T. Boothroyd, D.F. McMorrow, Phys. Rev. B **88**, 214415 (2013)
19. V.E. Dmitrienko, E.N. Ovchinnikova, S.P. Collins, G. Nisbet, G. Beutier, Y.O. Kvashnin, V.V. Mazurenko, A.I. Lichtenstein, M.I. Katsnelson, Nat. Phys. **10**, 202 (2014)
20. J. Herrero-Martín, A.N. Dobrynin, C. Mazzoli, P. Steadman, P. Bencok, R. Fan, A.A. Mukhin, V. Skumryev, J.L. García-Muñoz, Phys. Rev. B **91**, 220403 (2015)
21. C. Donnerer, M.C. Rahn, M.M. Sala, J.G. Vale, D. Pincini, J. Strempfer, M. Krisch, D. Prabhakaran, A.T. Boothroyd, D.F. McMorrow, Phys. Rev. Lett. **117**, 037201 (2016)
22. S.V. Grigoriev, V.A. Dyadkin, E.V. Moskvin, D. Lamago, T. Wolf, H. Eckerlebe, S.V. Maleyev, Phys. Rev. B **79**, 144417 (2009)
23. C. Pappas, E. Leliévre-Berna, P. Falus, P.M. Bentley, E. Moskvin, S. Grigoriev, P. Fouquet, B. Farago, Phys. Rev. Lett. **102**, 197202 (2009)
24. S.V. Grigoriev, E.V. Moskvin, V.A. Dyadkin, D. Lamago, T. Wolf, H. Eckerlebe, S.V. Maleyev, Phys. Rev. B **83**, 224411 (2011)
25. S.V. Grigoriev, N.M. Potapova, S.A. Siegfried, V.A. Dyadkin, E.V. Moskvin, V. Dmitriev, D. Menzel, C.D. Dewhurst, D. Chernyshov, R.A. Sadykov et al., Phys. Rev. Lett. **110**, 207201 (2013)
26. J. Kindervater, W. Häußler, M. Janoschek, C. Pfleiderer, P. Böni, M. Garst, Phys. Rev. B **89**, 180408(R) (2014)
27. S.L. Zhang, G. van der Laan, W.W. Wang, A.A. Haghighirad, T. Hesjedal, Phys. Rev. Lett. **120**, 227202 (2018)
28. B.T. Thole, G. van der Laan, Phys. Rev. B **38**, 3158 (1988)
29. G. van der Laan, A.I. Figueroa, Coord. Chem. Rev. **277–278**, 95 (2014)

Chapter 6
Dichroism Extinction Rule

In Chaps. 4 and 5, we have shown that for a modulated magnetic system, the circular dichroism at the resonant magnetic diffraction condition is extremely useful for retrieving the structural information of the motifs. As the chiral and topological nature of the motif structure closely relates to the CD extinction condition, we call this method the *Dichroism Extinction Rule* (DER). We will summarise this rule as a conclusion of the thesis.

The experimental conditions are:

1. REXS experiment.
2. The material has a spatially modulated spin structure. This can be directly confirmed in the same REXS experiment.
3. The motif of the modulation takes the form of, or can be approximated by, spin harmonics. This can be directly confirmed in the same REXS experiment where CD presents.

For a chiral magnetic system, the modulation wavevector is placed within the q_x-q_y-plane. Using circularly polarised incident light to reach the diffraction condition for that wavevector, the circular dichroism is defined as the diffraction intensity difference between the left- and right-circularly polarised light. Let CD probe the magnetic wavevectors from different azimuthal angles in a 2π range. Under certain azimuthal angles, the CD goes completely extinct.

- The number of the extinctions reveals the topological winding number of the motif.
- The sign of the slopes around the extinction conditions reveals the chirality of the modulation.
- The azimuthal angles of the extinction reveal the azimuthal canting angle of the spins within the x-y-plane with respect to the propagation wavevectors.

© Springer Nature Switzerland AG 2018
S. Zhang, *Chiral and Topological Nature of Magnetic Skyrmions*,
Springer Theses, https://doi.org/10.1007/978-3-319-98252-6_6

- The asymmetry of the profile reveals the titling angle of the spins within the x-z-plane with respect to the propagation wavevectors.
- The scale of the CD amplitude reveals the non-modulated spin component that tilts towards the propagation wavevector.

This concludes the full description of the dichroism extinction rule.

Curriculum Vitae

Shilei Zhang received his B.Eng. and M.Eng. degrees in Materials Physics from the University of Science and Technology Beijing. As a visiting scientist in University of Oxford's Clarendon Laboratory, he worked on magnetic devices, before joining in 2012 as a D.Phil. student in physics under the supervision of Prof. Thorsten Hesjedal.

© Springer Nature Switzerland AG 2018
S. Zhang, *Chiral and Topological Nature of Magnetic Skyrmions*,
Springer Theses, https://doi.org/10.1007/978-3-319-98252-6

Printed in the United States
By Bookmasters